## About the Author

Born in New York City in 1968, ARIEL LEVE grew up with her mother, a poet, in Manhattan. At the age of five, she began traveling to Southeast Asia, where she spent part of the year living in Bangkok, Thailand, with her father, a lawyer. She is an only child.

Leve has been a journalist and columnist with *The Sunday Times Magazine* in the UK since 2003. She has written more than a dozen cover stories, in-depth investigative features, and interviews. She has twice been nominated for the British Press Awards, for Interviewer of the Year (2005) and Feature Writer of the Year (2008), for which she was Highly Commended.

In 2008 she won Feature Writer of the Year from the Magazine Design and Journalism Awards. Since 2005 her column, "Cassandra," has appeared weekly in *The Sunday Times*, and prior to that it ran in *The Guardian* under the title "Half Empty."

Her work has appeared in Granta.com, *Vogue* (UK), *The Guardian*, *Elle*, *Marie Claire*, *The Evening Standard*, *The Jewish Chronicle*, *The New York Observer*, and other publications. She was one of the contributors to a collection of humor writing in America called *101 Damnations*. She is based in London and New York.

"Welcome to the wonderfully dark and witty world of Ariel Leve. This collection of her much-loved 'Cassandra' columns . . . is packed full of laugh-out-loud one-liners and thoughtful observations on everyday life. . . . In this forthright, funny, and extremely honest offering, Leve says the unspeakable that you'll doubtless have thought but felt instantly bad about and then tried to forget. Imagine the musings of an even more world-weary Enid from *Ghost World*, but for grown-ups." —*The List* (UK)

"Ariel Leve possesses humor that is both perceptive and caustic. If you enjoy laughing at your foibles and the foibles of your friends, then every line of this very clever, if painful, book will be chuckled and cried over." —*Sydney Morning Herald*

"The antithesis to Bridget Jones, Leve offers a refreshingly dark and wry outlook on life that is masterfully told through her sharp and acerbic wit." —*She* magazine

"Ariel Leve's dry and bitter wit highlights the funny side in our everyday negative feelings." —*Evening Herald*

"Darkly humorous Ariel Leve lays out her daily insecurities like a patchwork quilt of modern anxiety and inadequacy, sparkling with tiny appliquéd gems of self-deprecating cynicism." —*Good Housekeeping* (UK), "Best Books for September"

"Sharp, witty, and ruthlessly observed. Leve writes brilliantly—even, perhaps especially, when preparing for the worst." —*The Jewish Chronicle*

"Perfect for le crunch—who needs to be upbeat when negativity is this entertaining?" —*In Style* (UK)

"Positivity seems to be the byword for surviving the modern age. But do we have to be so bright and perky all the time, especially as the world economy crashes? Ariel Leve's dry and bitter wit highlights the funny side in our everyday negative feelings." —*Irish Herald*

IT COULD BE WORSE, YOU COULD BE ME

# IT COULD BE WORSE, YOU COULD BE ME

*Ariel Leve*

HARPER  PERENNIAL

NEW YORK • LONDON • TORONTO • SYDNEY • NEW DELHI • AUCKLAND

HARPER ● PERENNIAL

Versions of some of these pieces were first published in Ariel Leve's "Cassandra" column for the *Sunday Times Magazine* and on-line, and in her "Half Empty" column for the *Guardian*.

First published in Great Britain in 2009 by Portobello Books Ltd under the title *The Cassandra Chronicles*.

HarperCollins books may be purchased for educational, business, or sales promotional use. For information please write: Special Markets Department, HarperCollins Publishers, 10 East 53rd Street, New York, NY 10022.

FIRST U.S. EDITION

*Designed by Lindsay Nash*

Library of Congress Cataloging-in-Publication Data is available upon request.

ISBN 978-0-06-186459-9

10  11  12  13  14     RRD     10  9  8  7  6  5  4  3  2  1

For my mother and my father.

# Contents

People like to say it will all work out.
But what if it doesn't?

# Introduction: My Day

Sometimes after I wake up I will stand in my pyjamas in front of the mirror over the sink and watch myself getting older. I'll feel lucky. I'm not the sort of person who has to rely on my appearance to get by in life. Who cares what I look like? I'm the sort of person who relies on personality. But then it will hit me. *That's* what I have to rely on?

I'll examine my teeth and wonder what my life will be like in ten years. Will I still be standing in the same pyjamas over the same sink? Will I be drying my hands with the same towel when I'm fifty, when I'm sixty, when I'm seventy? People like to say it will all work out. But what if it doesn't? I will sit quietly on the edge of the bathtub and think about this for a while. Then, when I feel fully prepared to accept the uncertainties of life, I'll begin my day.

Worrying is my yoga.

There are two things in life I enjoy. Talking on the phone and drinking coffee. Better yet, talking on the phone while drinking coffee. That's about it.

There is no question I hate more than 'What's new?' Except maybe 'What's up?' There is only one answer I will ever give to those questions. Nothing. Nothing is new and nothing is up. Especially since yesterday.

The problem with leaving the house is occasionally I'll run into people I don't want to see. Recently I bumped into a friend I'd lost touch with since 1996. As we stood on the street corner waiting for the light to change, not having seen each other in over a decade, she asked what was new. At first I was exasperated by the absurdity of having to sum up over ten years of my life in five seconds until it hit me: I could do it. Not that much has changed.

What I'd like to know is, why does something always have to be new? What is all this new stuff going on? People seem to be experiencing new things all the time. Most of the new things I experience aren't worthy of reporting. Yesterday I found a new grey hair. I'm also concerned that I may have arthritis.

But I suppose what people are really asking is: is there anything going on in your life that will impress me. Let's see. I switched toothpaste, paid the phone bill, and bought a humidifier. I'm no Sharon Stone.

The first conversation of the day is usually with the deli man downstairs. I'll say, 'Good Morning.' He'll say, 'The usual?' I'll say, 'Yes.' Sometimes I might ask for an extra shot of espresso in my latte. After that, I feel energetic. That's new, I guess.

The rest of my day is an act of postponement. I'll sit at my desk before I begin to work and run through the list of worst-case scenarios until I get tired and can't think any more. Then I'll lie down.

Lying there I'll remind myself not to get too attached to things because you never know, anything can happen. It can all go away. Thinking about this – how nothing matters – can be a relief. As long as you don't think about it for too long. Then it becomes depressing.

When I'm done with work I might consider a visit to the gym. But the last time I was there I was on the treadmill listening to music when a trainer I had been avoiding came over the say hello. He stood in front of the machine I was on and smiled, forcing me to remove my headphones.

'Hey,' he said. 'What's up?'

I smiled. 'Not much.'

I decided this was more life-affirming than 'Nothing.' I try to be positive. Then just as I was about to put the headphones back on he made a face of disbelief.

'Come on. You're joking.'

How is that joking? I promised him I wasn't. I had nothing new to report.

'Sure you do,' he declared.

'No. Really. I don't.'

He looked confused. Suddenly, it was a standoff. He refused to accept my answer and I refused to adapt it to please him. Eventually he shrugged and walked away.

I watched as he approached a woman on the stair machine who I could tell was very proud of her midriff. 'What's going on?' he asked. Twenty minutes later they were still talking.

On my way home I noticed the streets were filled with people and I thought, what are they doing with their life that's so interesting? They go out, they shop for DVDs, they buy pomegranate juice. Each person that passed looked more miserable than the next. It was exhilarating.

When I got into the lift there was a woman with a hump. I could have a hump. Everyone has problems. I need to remember that.

I've never understood when someone says there are not enough hours in the day because from the moment I wake up the countdown begins. There are plenty of hours in the day. They just keep on coming, one after the other.

Mostly, I'm looking forward to turning ninety. Here's why. All the moments I've had to seize will be behind me. The pressure to have taken advantage of new opportunities will be off and just getting through the day will be considered an achievement. I'll have the same outlook I have now – but no-one will mind. And when someone asks me, 'What's up?' I can reply, 'Dinner.'

It will be enough.

# one

## GETTING THROUGH THE DAY

# Deli Man

Every morning I wake up and order a double latte from the deli downstairs.

In New York they deliver anything, any time. No matter what, the deli man always sounds delighted to hear from me. 'Hello, honey!' It's like a husband. Only better. I don't have to see him. And, more important, he doesn't see me.

All I am to him is a raspy voice that says slightly suggestive things like 'Can you make it extra-hot?' and 'I'd like a ripe banana too.' The conversation never lasts longer than three minutes and ends with certainty. We'll speak again tomorrow.

We have a special connection. I know him as 'Deli Man' and he knows me as '15-A'. Once, he asked for my name but he has since forgotten it. I don't care. As long as he gets me my extra-hot latte in under twenty minutes.

And I get preferential treatment, so I feel special. If I don't have enough cash, he'll let me pay the following day. If I tell him I've been up all night working, he'll throw in an extra espresso shot.

'You work too hard,' he says. I appreciate his concern.

He's intuitive too. He picks up on the fact that before I have my coffee, I don't really want a long conversation. If he asks me 'How are you?' I respond with 'Fine,' and that's enough. He doesn't push it. Sometimes, before I go away, I'll let him know I'll be out of town so that he doesn't think I'm abandoning him.

But instead of asking where I'm going and what I'm doing, he'll say simply: 'Okay, honey, have a good trip.'

Of course, when he has no time to talk to me, that's different. There are some mornings I can tell he's rushing me off the phone. I feel rejected and find myself dragging it out. 'Maybe I'll have an iced coffee today . . .' Then he'll snap at me: 'Come on, honey, I got others waiting here!'

That's when I realize we're not meant to be.

The best thing about my relationship with Deli Man is he had no idea what I look like. All I know about him is that he has a Mediterranean accent, whereas the delivery man is Mexican. I've often wondered if he reports back on what I'm wearing. For instance: 'She was wearing the plaid pyjamas again.'

I make sure to tip him well.

There are days I think it's the ideal relationship. I accept that he'll never be available on a Sunday, and I don't take it personally. I also enjoy the fact that he thinks I'm a lot more enthralling than I am. As 15-A, I'm a mystery.

But then, one morning, everything changed. I ordered the usual, the doorbell rang, I opened up and there he was. He'd delivered it himself.

Why would he do that? 'Hello, honey,' he said – but he looked so disappointed.

I was in my glasses and pyjamas, looking like an un-caffeinated, unshowered mess. What did he expect? Pamela Anderson? It was horrible.

There was only one thing to do. I pretended not to know it was him, handed over the money, grabbed my latte and shut the door.

We'll see if he calls me honey again. Or if he even answers the phone. But even if he sounds the same, I'll know that on the inside, his dream has died. We both lose out. He's lost the fantasy of talking to someone hot. I've lost the special attention that comes from him thinking I'm hot. And most troubling of all? I have to start making my own coffee.

# Something to Look Forward to

Everyone needs something to look forward to. Without it, there's nothing to keep us going. But there is a big drawback. As soon as it's happened, you're back to square one. Even as it's happening, you may begin to feel strange and wonder what that feeling is. I can tell you what it is. It's the void creeping back and reminding you you've got to come up with something else, fast, or you're going to crash.

All my life I've had to have something to look forward to, and it's a constant struggle to replenish. It's exhausting, like laundry. As soon as it's done, you're already into the next load. It used to be that I could look forward to growing up. I hung onto that for as long as possible.

Even though I had a suspicion it would be a let-down, I didn't really accept it until I was thirty. That's when it hit me: thirty more years to fill?

My best friend Liza always makes sure she has a date lined up. That gives her something to look forward to. Once she is on the date, it's another story. Then she can look forward to it ending.

Because my father lives so far away, I've had talking to him and seeing him to look forward to. So last week, I called him in Bali. I told him I was looking forward to our trip to Italy. He said he was too. I told him I was looking forward to us having some time together, face to face, to reconnect and talk.

'What are you looking forward to?' I asked. He said he was looking forward to getting off the phone.

Another friend, Audrey, is in love and all she can talk about is how much she's looking forward to the weekend, looking forward to the evening, looking forward to lunch. Today my lunch was a yoghurt and half a brownie. I looked forward to the brownie.

And then it was gone. Which is precisely the problem. Now what do I have?

My eyesight is deteriorating, my skin is only going to get blotchier, my metabolism is slowing down and my cholesterol is going up. I suppose I can look forward to my next blood test.

I tried to think of more things to get me through the day. I looked forward to there being only a few more hours before I could go to sleep. Oh, and my hair's getting longer. That cheered me up. Then I remembered as soon as that happens I'll have to book an appointment to get it cut. Not looking forward to that.

Recently, I was working on a story about someone who has just had his leg amputated. He said he was looking forward to getting a prosthetic leg and going horse-riding. You'd think that would put things in perspective — but it didn't. He has something to look forward to and I don't. I felt even worse.

I think that the whole idea of looking forward to something might be the problem. That's when I decided: from now on, I'll look forward to nothing. That way, I won't feel let down. I have nothing to look forward to.

I'm feeling better already.

# Napping

People are very judgmental when it comes to naps. Especially a nap in the middle of the day. If it's four o'clock in the afternoon and you tell someone you've just woken up from a nap, the disapproval hangs in the air. There is a self-righteous tone in their voice. Why? Because they were awake while you weren't. Here's what I want to know: why is it so important to live every second of life conscious?

Usually, a nap is for when you are tired. Not in my case. For me, the purpose of a nap is to get to the end of the day sooner. There is nothing better than waking up with the better part of the day behind me. Closer to dinner.

The problem is I have to justify it. So the excuses kick in. Jet lag is a great excuse. I love this one because afternoon, morning, early evening – it always works. Unfortunately, it has an expiration date. If you're still claiming jet lag three weeks after a trip, people get suspicious. Or, claiming jet lag from a train ride. They don't buy it.

'I was up all night' is good. But only if you add 'working' after it.

My friend Heather once broke up with her boyfriend because of his napping. She gets up at 5 a.m. to go to her high-powered job as a TV producer. He woke up at 10 to get to his menial job that required no skills. She would call before lunch to see what he was up to and he would say: I'm taking a nap. He wasn't busy enough to be napping.

'You have to earn a nap!' She says, still annoyed. 'What was he tired from? Breakfast?'

Napping is a complex subject. It's not only why or when you nap, but where. Another friend, Lori, is an afternoon napper who's taken to hiding it from her husband. If she tells him she just took a nap he'll ask: where were you? He doesn't mind if she naps on the sofa with the sun streaming in. But if she naps on the bed, it's an issue. He finds it depressing.

I've tried to figure out what this is about and I think it's because a nap on the couch is accidental. It says: I didn't mean to fall asleep; I must have just drifted off. But a nap on the bed is definitive. It says: I made a commitment. I intended to sleep the day away.

I only nap on the bed.


# Medically Induced Coma

Sometimes I wish I could be placed in a medically induced coma. Is that wrong? Just for a few days. I'd wake up feeling rested and refreshed. I'd lose a few pounds and the dark circles would be gone. It would be like having time off.

Maybe a few days isn't enough. Now I think about it, I wouldn't mind being comatose for the month of December. Every year. I could avoid all the holidays, then on New Year's Day I'd wake up and ask if anything new had happened while I'd been out. Had anyone paid my bills for me? Did my apartment clean itself? Was my neighbour evicted? No? Okay, put me down for another month.

It would be a good way to save money too. It's impossible to get through a single day without a trip to the cash machine, but in a coma I could stick to a budget. I wouldn't be using my mobile phone and what I'd save on groceries and shoes I'd put towards medical procedures. I could have them all done at once – my fillings replaced, my eyes lasered. I'd wake up with an absence of tooth decay and 20/20 vision.

But I'd need to be groomed. Liza would have to get someone to come in and tweeze my eyebrows, give me a mani-pedi, and maybe some highlights. I'm far too impatient to sit still at a salon, but in a coma there's nothing but time.

I'd probably want to keep it going. Fifty years in a coma doesn't sound so bad to me. I'd wake up just in time to die.

The ideal way to be in a medically induced coma is when I'm waiting to hear back about something. For instance, let's say my novel has been sent to the publishers. That's a time when I'll be filled with anxiety and unable to focus on anything else. So my instructions would be: wake me up when there's a book deal.

Another good time for a coma: just after getting dumped. People say that time heals all wounds. Well, time passes when you're unconscious, so why not fast-forward to six months later, fully healed.

For some people, the best part of being in a coma has got to be waking up. When you open your eyes, you're surrounded by friends and family who are all crying tears of joy at your recovery. But if I was in a coma, who would hold a vigil at my bedside? Thinking about this, I could feel myself getting depressed.

I don't want to wake up from a coma alone. I could see myself shouting: 'Hello? I'm out of my coma now? Anyone there?' I wouldn't know what to do.

The other day I called my GP, Dr Samuelson. He told me a coma is only induced when a patient is in dire circumstances. 'Define dire,' I said. 'When the brain needs to rest,' he replied.

I told him my brain needed a rest. There was a long silence.

Then he hung up.

# Complaining vs Lamenting

Complaining can be about anything and everything, day or night, anywhere, anytime. Complaining is a lot like talking, only more constructive.

People often confuse telling a story with complaining. But every good story needs a complaint. Otherwise, what's the point? Who wants to hear a story about how great someone's life is? If I'm going to participate in the discussion, there has to be something in it I can relate to. A complaint makes it interesting.

A lament is far more complex. I consider it a sophisticated version of complaining with weeping, despair and regret mixed in. For instance, I would lament the fact that I'm in a tiny apartment with a weird smell of pickles and chicken and now all my clothes smell of pickles and chicken and the dry cleaner ruined my sweater when I sent it in to have the smell removed and soon it will be the holidays and I'll be alone and what if it never changes, in spite of what people say, and this is all there is. I complain about the elevator.

Recently in the elevator, I found myself enjoying a conversation with another tenant in the building. This rarely happens. People always feel compelled to talk to me and usually, that's what I'm complaining about. But yesterday was different. Why? Because we were

complaining in tandem. She complained about how slow they run –
then I complained about how one is always broken – then she nodded
and complained about how complaining to the management is useless,
and I heartily agreed and complained that it's the same with the laundry
room and the broken washing machines. Next thing I knew, the
sixteenth floor had arrived.

There are three elevators in the building but one is continually out
of service, so waiting requires patience and stamina. I have often found
my neighbour in the hallway with reading material. Even if he's only
going downstairs to collect his mail, he knows he'll be there for a while
and passes the time productively.

I, on the other hand, prefer to stand idly by getting older and more
anxious with every passing second. I tend to use the time to ruminate
about my life and itemize the bad choices, putting them into categories
like Work, Men, General, and so on. I don't want anyone interrupting
me. Unless it's with a complaint.

# A Carrot for One

I was in the supermarket, waiting in line to pay for my carrot. An elderly woman was in front of me holding a lime. We looked at each
other and smiled. That's when I realized: I'm ninety. When you're elderly, you don't mind waiting in line for twenty minutes. What's the
rush? Then the elderly woman looked at her watch.

Why would she do that? Maybe there was somewhere else she had
to be. Someone waiting for her to get home with the lime? Or a party
she had to get to? This bothered me. She had somewhere to be. It ruined
the moment.

Usually I enjoy standing in line quietly. It's where I do my best
thinking about where it all went wrong. Standing in line is acceptable

time out. Sometimes I like to think about other things, like why the line I'm in isn't moving and the one next to me is – but I never feel like switching because if a line isn't moving, that's the one I want to be in.

The last time I went out to dinner with a friend she put her mobile on the table. So I put my phone on it too. Hers rang. Mine didn't. Hers rang again. Mine didn't. I put mine away. The only thing worse than a mobile on the table that rings is a mobile on the table that doesn't.

Liza is never without her BlackBerry. This summer, I was in a car with her for three hours stuck in traffic. Something I didn't mind. Except, the entire time, she was working. Between making deals and her social life, it was nonstop. 'I can't do Tuesday: I've got a book party at six and an opening after. Wednesday I can do . . . after the premiere.' It was awful. Who wants to hear that? At one point I was so excited because my phone rang. It was my father.

I've always wondered what it would feel like to be someone who has so many people needing their attention that they complain about it all the time. There are people who say, 'Everyone wants a piece of me,' and whenever I hear that I think: 'What on earth are you talking about?' No-one wants a piece of me.

I bet if I was the type of person who had a BlackBerry or a tan or, better yet, a BlackBerry *and* a tan, I'd be eager for whatever line I was standing in to move quicker.

But then I'd start thinking how much I missed the time in my life when I had twenty minutes to spare.

After I purchased my carrot, I was walking home when I was confronted by a blockade of people walking three abreast. How is someone meant to get past three people taking up the entire pavement? 'Excuse me!' I shouted. Then I elbowed past them. I might not have had anywhere I needed to be, but I had to get there as soon as possible.

# Coffee

People are always doing studies. Now there's one that says drinking coffee can lead to the prevention of memory loss in old age. This is terrible news. Drinking coffee is my greatest pleasure in life. That, and forgetting.

I first started drinking coffee at the age of seven. Even at that age I knew I wouldn't make it through the day without an incentive to get out of bed. Every morning it was the same: white coffee, two sugars. You know those memory games that children play in the car? I was always winning. Now I know why.

One day at lunch my science teacher saw me drinking coffee in the cafeteria and complained to my mother. He told her that he didn't think it was a good idea for me to have caffeine and that's probably why I wasn't paying attention in class. But she told him the reason I didn't pay attention in class was because he was a moron. After that, he left me alone.

Now that I've learnt coffee improves my memory, I'll have to reconsider how much of it I drink.

During my last relationship I would recount entire conversations — sometimes from two years earlier. I'd point out something that was an inconsistency and feel proud as my ex looked stunned. I've found there's nothing a guy loves more than to hear a woman begin a sentence with: 'But you said . . .'

The main problem with having an acute memory is, why would I want to spend my old age recalling parts of my life that while they're happening, I'm not really engaging in? That's why I never take photos. Everyone is so excited about digital cameras but I don't get it at all. I'm not that into reliving a moment if I didn't enjoy it the first time round.

Not to mention that digital cameras aren't really capturing the moment anyway. They're capturing a moment after you've made sure you don't look like a hippo.

When I was in Italy over the summer, every morning I would wake up and all I could think about was that soon I would be drinking a double espresso. The time it took between my eyes opening and getting to the café in the square where I could be caffeinated was time that did not exist. It was a fifteen-minute interim during which I managed to brush my teeth, splash some water on my face, get dressed and trudge up the hill. No talking. Once I got to the café, I'd order the espresso with hot milk on the side. The only Italian word I remember is *schiuma*, which means foam.

I'm inclined to keep drinking coffee and suffer the consequences later on. Here's my plan. Just when it gets to the point where all the memories start taking over, I'll quit. That way, the cold-turkey migraine headaches and sleeplessness will divert my attention.

I can't imagine a life without coffee. The way some people can't imagine a life without children.

# Therapy or Dinner

Lately, I've been thinking I should go back to therapy. The problem is, I live in the West Village and my therapist's office is seventy blocks uptown on the Upper West Side. It's such a commitment. Leaving the apartment, walking to the subway, waiting for the train to arrive, possibly not getting the express then walking another three long blocks, in the rain. I should be more motivated.

My friend Audrey is the same way. She's shopping for a shrink at the moment and her only criterion is that he or she is in the same neighbourhood. 'I loved my old shrink,' she says. 'But I had stop going because her office was across town.'

This is someone who would cross the Alps barefoot for a sale at Diane von Furstenberg. A few weeks ago she went to a sale at Frette linens in DUMBO. Brooklyn. A different borough. When it comes to

Egyptian cotton sheets, she's audacious. When it comes to mental health, not so much.

Money is also an issue. My therapy isn't covered on insurance so for the past few years I've had to choose between dinner out or a healthy future. Turns out I'd sooner spend $75 on a piece of salmon than sorting out my emotional wellbeing.

That's not a good sign.

# two

## FRIENDSHIP

# The Loser Friend

There is tremendous value in having a loser friend. Someone far worse off than you who is slightly in awe and worshipful of your life. You complain about this friend constantly and your other friends get annoyed and say, 'I don't get it. Why are you still friends with that person?'

But you can't cut them loose because you need to have someone there to remind you it could be worse – you could be them.

The loser friend creates conflict. Conflict is good. It's in every good book or movie – and what people don't realize is, life is no fun without it. Having the loser friend keeps things on an even keel. It gives you something to complain about – but still reinforces your self-esteem.

My loser friend – Carl – is an even bigger mess than I am. I avoid his calls as much as possible because I know once we're on the phone, it will be at least an hour before I get a word in. I'm all for obsession but in his case, it's a bit much. He's delusional and obsessed with a girl named Claire and he says things like: 'I know she loves me, she just doesn't know it yet.'

This has been going on for years. And for years, I've been complaining about him – how I can't stand it – how he drives me nuts – how he doesn't listen. Then, just when I think I'm on the verge of ending it, he'll say: 'You're so right. You're SO right. I wish I was more like you.' And suddenly, he's not that bad.

The loser friend is generally someone who is very competitive with you and thinks they're on an equal footing. Carl will do or say something utterly pathetic and appalling. Then he'll sigh and state: 'But you know what I mean. We're cut from the same cloth, you and I.'

The problem is, if I cut him loose, it's worse. He could find dignity and be above talking to me. As long as I keep him in my orbit, I have the upper hand. If he were out there – there's always the chance he could recognize that *I* was the real loser for having kept him as the loser friend.

# You Told Me Already

Whenever a close friend begins a conversation with, 'Tell me if I've told you this already,' I know nothing but trouble lies ahead. The sentence suggests a confidence or a conspiracy is about to be shared, a sexual faux pas, a health issue, a disaster that's befallen someone we both loathe – but it immediately raises suspicion too.

If you can't remember telling me your secret in the first place, you must be telling the same story to others, so I no longer feel like a confidante. And if it's a story you've told me before and can't remember, I'm clearly not special enough that you'd recall sharing it with me originally. If it's a story you haven't told me, I'll wonder why you were worried you might have, and suspect you told it to someone else, which means I'm not the only one you trust, so my hackles are rising even before the story begins.

The problem for people who repeat stories is that most listeners are too polite to point it out. People will endure hearing a story two or three times because they don't want to embarrass the person telling it. Not me. I'll say, 'You've already told me this,' and I won't hide the irritation in my voice.

What happens next is denial. They'll make a face of disbelief and say: 'Are you sure?' Then I have to repeat their story back to them until they believe I'm not lying. Even then, sometimes they're not convinced. 'What about the part when . . .' Yes. You told me. I have heard the story. I promise you.

Nobody wants to be reminded they've told the same story twice, because it makes them seem flaky. But shouldn't they have this information so they can do something about it?

Liza and I had a fight about this. She began to tell me a story she'd told me two days earlier. Two days. How is that possible? 'You don't remember?' I asked.

No, she didn't. And then she was the one who seemed annoyed. Why? 'Because,' she said, 'you sound so hurt. You take it so personally. You'd think I'd forgotten you were the person I'd lost my virginity to.'

How could I not take it personally? If someone is telling me a story again, this says that what I thought was a bonding moment between us was merely a widespread bulletin that went out to friends, co-workers and family. Of course, it depends on what's being repeated and who's doing the repeating. If my doctor is about to repeat something, I don't mind. He's got hundreds of patients, I don't see him that often, he's off the hook. Plus I'm always willing to hear a medical cautionary tale. But with friends and in romantic relationships, it's different.

'If it makes you feel better,' Liza said, 'you're not the only person I repeat stories to.'

That did not make me feel better. It only confirmed what I'd feared. But it got me thinking. Maybe it's me. I have far fewer people in my life than she does; I don't find it difficult to remember who I told what to, because I don't have that much to tell and few people to share it with. And now I'm wondering – if people forget the stories they've told me, do those same people forget the stories I've told them? I hope so. Most of the stuff I share I regret afterwards, so it's reassuring to think they'll forget they heard it in the first place.

Chances are nobody's listening anyway.

# Sophie

I won't call it a love affair because that suggests there will be an end. It's more like a love story. Best of all, there is no snoring, no post-coital regret, no unreturned phone calls or awkward silences. From the moment we met we knew there was chemistry and since then, not a day has gone by where we haven't spoken.

I have made a new friend for life. Her name is Sophie and we met through an ex. A few years ago I briefly dated Jack, the sort of man who does not have 'girlfriends'. He is an ex-something and we remained friends. Sophie is his new something.

Whenever Jack spoke about Sophie he would use her name. I knew this meant she was special. Usually it's 'The one from Georgia' or 'The go-go dancer' or 'The clutcher.'

He also mentioned how similar we are. Both of us writers, with challenging childhoods and Jewish. Or as Jack put it: we both love to talk.

The three of us went out to dinner. Instantly it went from 'nice to meet you' to finishing each other's sentences. When I got home that night there was an e-mail from her waiting in my inbox. I wrote back immediately. The following morning, the discussion continued over e-mail and then I can't remember who called who first but we stayed on the phone until we made plans to see a movie later that afternoon and then after that we talked on the phone again into the evening.

It was so easy and uncomplicated. I didn't worry if I was calling too soon. I didn't wonder if I was making myself too available or if maybe I'd imagined the connection or if she'd change her mind and never call back. Instead the conversation flowed seamlessly and every topic – from having friends in common, being born two days apart or liking the same cookie from the wheat-free bakery – seemed exciting.

Plus, we had the same taste in men. That's five years of conversation right there.

It's hard to make close friends as you get older. People come in and out of your life but when you meet someone you don't have to explain yourself to — it's rare. And this was the ideal relationship — the kind where you call and say, 'It's me' and the person at the other end doesn't reply, 'Me who?'

On Monday I called Jack to thank him. In twenty-four hours Sophie and I had probably had more verbal communication than Jack would have in his lifetime.

I told him I would always be grateful he'd introduced me to her because it felt like I'd met my long-lost sister. Being separated at birth means there's a lot of catching up to do.

Just then, as we were on the phone he said an e-mail arrived. It was from Sophie. She was writing to thank him and said she felt like she'd met her long-lost sister. He was getting the gratitude in stereo. 'She said that?' I asked. 'She used those same words — long-lost sister?'

We use the same words and feel the same way. If only she were a man — I'd be done.

Then again if she were a man, it probably wouldn't work out. I'd never be able to trust it because it would be too good to be true. And no amount of reassurance would help. Soon it would sour. Anticipation would turn to dread. I'd call but days would go by before I heard back. *If* I heard back. Then I'd feel rejected. And angry. I'd lose weight.

Thankfully, that's not going to happen. In fact, the only thing that troubles me is that I can't see a downside. I'm sure I have the capacity to screw it up — but how? It's worrying me that I can't see this. It means I can't prevent it.

# Facebook

I know most people see Facebook as a positive thing and an opportunity to make new friendships, but I'm worried. It's opened up a whole new world of paranoia. I always knew I had the potential to alienate people in real life but now I can drive away thousands of people in cyberspace too?

I joined Facebook because I was told to I had to. 'You're missing out,' Sophie said.

I enjoy missing out. There's plenty to be gained from it. Whenever I've panicked about missing out and then forced myself to go somewhere, I've invariably returned home thinking: I wasn't missing much at all.

But missing out on what's going on with my friends is different. The disparate nature of people's lives means I feel like I'm always in catch-up mode even with close friends.

So Facebook is a tool for staying in touch. For most of my friends it has replaced e-mail. Now even texting and phone calls have become a chore. I'll get through to Madonna before I'll get through to Liza.

Plus, I've always wondered what my friends are up to when they're not available to meet. Now I know. Facebook means I can get status updates and as soon as I joined, one instantly appeared on my screen.

'Liza is folding laundry.' I felt so included. God only knows what else I've been missing.

The messages poured in congratulating me on joining. One said: 'I can't believe it!' and another: 'I'm so happy for you.' Then the phone rang. It was Sophie. 'I'm so excited!' she said. 'Finally!'

You'd think I'd gotten engaged.

In real life, my friends are uninterested and distracted. But in cyber life people are very excited (!!!) about everything!!! The levels of emotion are off the charts. Everyone is so effusive. And curious. People I've never met before and have no interest in meeting are asking me questions about what I think and how I feel. My new fake friends have raised the bar.

A lot of people who have become friends with me have no idea who I am. Which suits me fine. These are great friends to have. I've also enjoyed reconnecting with people I haven't been in touch with for twenty years and catching up in two sentences. Does it get any better? For the most part we're saying: 'Hi. Not dead.'

The only thing better than having a social life without ever leaving the house is having no obligations or responsibilities for maintaining this beyond an occasional click of a button.

You can't get mad at a friend on Facebook for not getting back to you and I've found that disappointment and rejection that would normally upset me is non-existent in cyber life.

All of which has led to a startling discovery: I'm so much better as a virtual friend than I am as a real friend.

The question is, how to carry it over?

Thanks to the status update I can read about what everyone is doing if they choose to post it. Once you've agreed to be friends with someone this update comes automatically. I thought about posting my own status update but figured 'I just woke up – going back to bed – may snore' wouldn't be of much interest.

I was wrong. There's no minutia too small to report. Nancy is tired. Karen is eating cereal. Then people can comment on these updates. I try not to waste time but it's hard to overlook comments like: 'What kind of cereal? Bran? I love bran!'

I wish I had that much enthusiasm about something.

Then there is information about my friends that is slightly unnerving. For instance, I'm waiting for Emily to get back to me and a week has passed. Suddenly, a status update appears.

'Emily has just spent eight hours doing nothing and is incredibly bored.'

Now what? I'd like to think I haven't heard back from her because she's busy with work or family commitments. But apparently doing nothing is preferable to talking to me. I don't want to know that.

I don't mind the status updates when people are blowing their nose.

I feel productive when I read those. But the ones like 'Mike is thrilled with his new multi-million-dollar deal' are trying. It's a constant struggle not to get depressed. And I don't even know who Mike is.

But the saddest thing of all is the people I used to be friends with, with whom I still share mutual friends. Facebook will voluntarily suggest the rejected friends as new friends. This means their photo pops up on my screen whether I want it to or not. This includes ex-boyfriends. And they're always smiling.

And what about the people I regard as friends in real life who haven't approved me yet? Are they ignoring me? There's no way to know. If I run into them in real life, do I pretend we don't know the truth? No one will ever say, 'Yes, I'm ignoring you' in person.

My friend Jon hasn't gotten back to me yet and I went to his page. He has 2000 friends. 2000. But I'm not worthy?

Conversely, the opposite isn't any better. I was approved by Rocco DeSpirito, the famous chef who I met once. That's sweet, I thought, he remembers. Until I discovered Rocco has nearly 5000 friends. Now I don't feel so special.

I predict it will all end badly. Sophie is annoyed because I'm not using the status updates. 'That's the best part!' Other friends have wondered why I haven't been in touch. I'll end up losing my real friends because I don't have enough time for them, thanks to keeping up with new people I'll never see or speak to. It's a lot of pressure.

# Breast Sharing

There's a new trend in motherhood. It's no longer enough to breast-feed your own baby, now you need to feed other people's too. It's called 'milk sharing'.

There's also an online resource where milk-needy mums can meet donors. It's like Facebook. But instead of photos, you're sharing breasts.

I don't have children so this is not something I need. But let's say I did. What's wrong with formula? Or a milk bank? There's a reason the wet-nurse is no longer a popular profession in places like the United States and Britain. There are options.

Besides, I can't imagine any of my friends being willing to breast-share. Forget hygiene, the major hurdle is, it's too much to organize. Between texting and e-mail, it takes days to co-ordinate a phone call. People are busy. By the time I got them to commit to a breast-feeding date, my baby would be in kindergarten.

Then they'd forget they promised to help out. Or their breasts would be too sore. Or they'd be too tired.

Except maybe Sophie. She's been thinking about getting her 34D breasts reduced – they give her back pain – so this could save her some money on the operation.

For the most part, my friends are not a charitable bunch. Asking to borrow a top is a major production. And I can't remember the last time anyone suggested sharing a bra. So lending a breast seems like a long shot. Not to mention the quid pro quo factor.

My friends keep score. If I ask someone to meet me near my house for dinner, you can bet next time, they'll point it out and use it to wring a concession from me. It's one thing to ask them to the Village. But to breast-feed my baby? It would give them the upper hand for life.

I can hear it now. 'I lent you my breast three times last month. You owe me.'

Of course I wouldn't want to help them out either. I'm an only child so I'm not good at sharing. I don't like sharing a toothbrush or popcorn at the cinema. Babies put things in their mouth that have been on the ground. How do I know where someone else's baby has been?

What's more, if my friend is asking me to breast-feed, in an emergency, I'd be suspicious of her reason. What if she considers getting a spa treatment an emergency?

I understand some mothers don't produce enough breast milk but the article says this cross-nursing trend is by choice — when mothers believe there is no substitute for breast-feeding, they turn to friends. It cites an example of a mother who left her baby with a friend to go to a job interview. The interview ran over and the breast-milk she left ran out. The friend used her own breast to feed the baby.

So let's say someone got stuck at yoga or in Barneys. Would they call their friend and say, 'Will you breast-feed for me? I can't decide whether to buy these Prada wedges.' Where do you draw the line?

I know some mothers are worried their baby wouldn't bond to another breast. Not me. I'd be too afraid of the opposite.

But also I can see this ruining a lot of friendships. None of my friends would agree to breast-share because they'd be too worried I'd micro-manage them. And if they're too tentative to ask me to dog-sit, I'm not going to be the person they think of asking to feed their baby.

I don't believe breast-fed babies are better off, anyway. I know they're supposed to grow up healthier, wiser, with a stronger immune system and higher self-esteem. But it can't always be true. I was breast-fed, and look how that turned out.

# Kidney Donor

Every other week it seems there's a new story about kidney disease. The latest was about two best friends — one who needed a kidney, the other who donated. That's got to put a strain on a friendship.

Naturally it made me wonder what would happen if I found myself in that situation. It wouldn't be easy. I have a hard time getting my friends to come downtown for sushi — I can't see anyone willing to give up a kidney.

'If it were a matter of life and death, I would,' said Liza. Then she added: 'But if it were because you felt sick or you thought it had just gone bad – I would say no.'

I found it touching that she was considering it. Even though she was essentially saying she didn't trust me to distinguish between a terminal illness and a lower backache. I thanked her and told her I'd take that as a yes. She paused. 'Let's talk about it later,' she said.

I called my friend Lori. She was far more definitive. 'Nope,' she declared instantly. 'What if my husband needs one? I'm saving it.'

I understood. At least she was saving it for her spouse. It would be awful if she'd said she was saving it for someone else and when I asked who she replied, 'The doorman.'

After talking to Lori I could definitely see why getting married has its perks. Built in kidney-donor. If I ever get engaged I'm making sure my fiancé has the right blood. From now on when someone asks what my type is I'm going to respond: O negative.

My friend Audrey was another non-committer. 'Maybe,' she said. 'It depends on my family.'

That's a good one. Of course I asked how many people are in the family. 'Small,' she said, trying to sound positive. 'My sister, my brother-in-law, my niece and my nephew.' That doesn't sound so small to me. Plus, she's including an in-law she doesn't even like. Second on the waiting list is promising. But fifth?

I moved on. It was not looking good. One friend told me I wouldn't want her kidney because it's not in great shape and another friend never got back to me. Simon, my British friend, was very polite. He would give me a kidney if it stopped me from dying, but he'd need to know more. Like what? A dialysis schedule. My ex-boyfriend told me he didn't think we were a match.

Even my own father was reluctant. I sent him an e-mail and at first he said, 'Anything to make you happy.' But seconds later he wrote back, 'Hang on . . . a kidney? Let me think about it.'

How is it that no-one I'm close to would give me a kidney? Would I have to go to my new Facebook friends for support? And what if no-one online responded? I can see it now – the status update says: Looking for kidney donor. The comments: zero.

Not that I blame people. I'm not up for donating either. And not because I'm saving it for anyone. Or because my kidney isn't working properly. Or because I'm worried about the aftermath. It's the potential resentment I'd have to live with. I'm worried that whomever I gave it to would do something to upset me and then I'd want it back. I'd be stuck. I'd end up saying something like, 'I gave up a vital organ for you and you can't return an e-mail?'

I don't want to be that person.

Then, just as I was resigned to accept my fate of life on a transplant list, Sophie called. 'I'd give you a kidney,' she said. 'Why not?'

Really? I was thrilled. She explained she was pretty ambivalent about life at the moment – so she didn't mind. I took that as a magnanimous gesture.

In fact I was so grateful I mentioned that if she ever felt enthusiastic about life again and changed her mind, I wouldn't hold it against her. It's the thought that counts. I probably wouldn't want a new kidney anyway.

## Audrey's New Nose

My friend Audrey is good about answering e-mails. When a few weeks went by and I hadn't heard from her, I worried. Then one day she wrote back: 'I'll call you, need to tell you something.' Naturally I wondered what I'd done wrong.

I was nervous, convinced she was ending the friendship. She began with an explanation: 'I know this will come as a shock, and the reason I didn't say anything before is because I didn't want you to talk me out of it.'

Was she getting married? Moving to LA? Or even worse: getting married and moving to LA?

She continued. 'It's over now and I need you to be supportive because there's nothing I can do to change it.'

I told her she was scaring me. 'It's nothing bad.' She paused. 'I've had some work done.'

Some work done. It sounded so Joan Collins.

Are we at that age now? I couldn't imagine what she'd had fixed. We've been friends for years. Close friends. We've travelled together, lived together, micro-analysed every insecurity, lump, bump, blemish. We've had plenty of discussions about plastic surgery and there's nothing she'd ever mentioned wanting to change.

When she told me it was a nose job, my first reaction was to feel left out. All those years she'd been keeping this lack of confidence a secret. How could she not tell me she hated her nose? If there was anyone who'd understand not being happy, it was me. If only she had confided in me. We could have been miserable together.

Then something occurred to me. The last conversation we'd had, I'd asked her to look at my nose and tell me if she thought it looked bulbous. I'd gone online to a rosacea website, seen a photo of W. C. Fields, and discovered there was something called 'W. C. Fields nose'. Was I getting that?

'That conversation was tough,' she said. 'I wanted to tell you but I wasn't ready.' She felt guilty. We had been on the nose topic and she hadn't said a thing. 'That whole time, I was looking at you, wishing I had your nose.'

Hang on. My nose? She was envious of my nose? I couldn't believe it. It was like she'd been having an affair under my nose, with my nose.

That changed everything. She'd never envied anything about me before. Nobody has. Of course, if there's something about me that's enviable, it's bound to be deteriorating. From that moment on, I've been on W. C. Fields-nose watch.

And then there was the anxiety about Audrey changing. What if once she had her new nose, she wasn't self-deprecating any more? What if she starts going out all the time, feeling on top of the world? That would be horrible.

I'd have nobody to talk to any more about not having a social life. Then again, she wasn't exactly a wallflower before.

I haven't seen the new nose yet, but I understand why she did it. If it makes her feel better about herself, why not? And she's lucky. There is a procedure available to her that allows her to improve.

For me, it's not so easy. It's not that there aren't things I wish I could change; it's that there are too many to count. I'm all for getting work done but doctors have yet to offer a personality transplant. When that's available, sign me up.

# Suicide Pact

I heard about a woman who made a suicide pact with two of her friends. If she ever became terminally ill, they agreed to assist her so that she could determine when she died, and do it with dignity. My first thought when I heard this was: she found two friends?

I'd be lucky if I could find one. Who could I count on to help me die?

As usual, the first person I turned to was Liza. She was at work. Here's what she said: 'Sorry, honey. I don't have time to give this any thought at the moment.' Not a good sign. If Liza's too busy to even discuss a suicide pact, what are the chances she'll have time to carry it out? Everyone in New York is so busy. But maybe I caught her at a bad moment.

I called again after work. 'If I said yes,' she said, 'you would ask me so many freaking questions about how I was going to do it and if it would hurt and if it would smell and if it would take too long, et cetera, that I would end up wanting to kill myself.'

I took that as a no. What kind of best friend won't agree to assist a suicide?

I asked Heather because she is reliable. I knew when the time came she would show up. That was my main concern. I didn't want to be left waiting, wondering if she had forgotten. Audrey would fit me into her schedule no matter what was going on. But she, too, was reluctant.

'Only if it was really terminal,' she said. She was concerned: what if I told her it was terminal — but it was only a bad hair day? What if I thought it was terminal but then it turned out to be curable? Who would make the decision when it was time to go? So many things to think about. It was getting too complicated. After a while she agreed, but only on the condition that I left her some of my furniture.

My furniture? What did she have her eye on? She told me she liked the art-deco dressing table. I panicked. I gave that piece away to Emily last month.

I called Emily. Would she consider giving me my dressing table back, and if not, would she consider helping me die?

She said she'd consider the latter. But on one condition. Everyone has their conditions. Emily wanted to know if she could have my apartment. Fine. But even after all that, she still wasn't sure. 'Let me think about it,' she said. She'd get back to me.

I was getting depressed. Nobody wanted to commit. And no-one was getting back to me either. I called my friend Linda in LA. But she was worried she wouldn't know what to do when the time came. What's to know? Just give me a bunch of pills. Or if there's a plug, pull it.

This was not going well. Nobody would agree to kill me. Not one friend. What does that say? That I need some new friends. Eventually Emily got back to me. She said she'd agree. Finally. 'You promise?' She promised. 'But I just want you to know,' she said, 'I'd feel guilty.'

That's okay. I can live with that.

# three

## IT COULD BE WORSE

# In Case of Emergency

There are people who are prepared for emergencies. And then there are people like me. The only crises I've ever encountered are emotional. I know what to do if someone is having a nervous breakdown. But if someone were having a heart attack? They'd die.

Not only am I unprepared, but my survival instincts are defective. Recently I was driving alone on an icy road when the car started to skid. As I lost control all that went through my head was that I'd like some hot ginger tea.

Luckily, there were no other cars on the road and when I got to a ditch, the car stopped.

Then, the next week, there was a fire in my kitchen. My toaster burst into flames a foot high. Since I was on the phone when it happened I did what my instincts told me to do. I stayed on the phone.

'There's a fire in my kitchen!' I shouted. To which my polite British friend responded: 'Oh, no.'

Just then, the fire alarm went off. 'Put it out,' she advised.

I was in a panic. To get past the toaster to the plug socket I'd have to put my hand through the flames. I knew enough to know that wasn't a good idea. I grabbed a bottle of water and poured it onto the toaster.

The fireman told me this was the worst thing I could have done. 'You could have been killed by the arc,' he said.

What arc? No-one told me about an arc. I never took a cooking class and it's not as though it would come up in a conversation with my doorman. Also, even though I Google illnesses frequently and can diagnose as well as any doctor, I haven't moved into life-threatening situations that don't have to do with disease.

Here's what else no-one told me. Toasting a rice cake is inadvisable. Even though it says on the packaging that it can be toasted, it's essentially like toasting a tissue.

But now that I was a survivor, I wanted people to know. How would my friends greet the news of my near-death experience? I called them and announced: I nearly died.

Reactions ranged from scorn to indifference.

Simon in London started laughing and called me stupid. 'You threw water in the toaster?' He couldn't get over my ignorance.

I explained that I didn't know you weren't supposed to do that.

'Everyone knows that,' he snapped.

He sounded so annoyed. Maybe if I'd died it would have been less frustrating.

I moved on and called Liza.

'Sounds bad,' she said. She was walking down the street with her friend Mary on their way to lunch. I told her I survived. 'That's great, honey. Mary says Hi. Gotta go.'

'Tell Mary I survived,' I said. She had already hung up.

I called my friend Dave who was also at lunch. Who knew so many of my friends went out to lunch? I explained what happened and waited for his response. 'Let me call you back,' he said, 'I'm about to order.'

Only Sophie had genuine sympathy and recognized the gravity of what could have been. 'I'm very glad you weren't electrocuted,' she said.

'Thank you,' I replied. 'Me too.' Then there was a pause. There's not much left to say after that.

I went out to buy a fire extinguisher. The man at the hardware store took one look at me and handed me one which looked like a bottle of hairspray. 'It's very easy,' he said, 'just pull the top off and press down.' I guess I must look like someone who wouldn't know what to do with it.

Just then it hit me. A fire extinguisher makes an excellent gift.

Birthdays, weddings, engagements, baby showers – finally I know what to give.

At first, no-one will think it's a good gift but you know when they'll thank me? Next time they have a fire. Then they'll appreciate it. I might not know how to save my own life but I can get credit for saving others.

# Sharks

It's been a bad time for sharks. A British tourist nearly got eaten off the coast of South Africa – barely escaped with his life. A strapping Australian got bitten in half. Then a teenager was attacked while he and his brother were fishing waist-deep in the Gulf of Mexico, sixty feet from the shore. Two days earlier a fourteen-year-old girl had been killed by a bull shark at another Florida Panhandle beach.

It's one thing if you're on a raft in the ocean in the middle of the night, naked. After we saw *Jaws*, we knew enough not to do this. But waist-deep in water? Sharks have gotten so bold they don't even try to hide themselves any more. The only reason to be on a beach is to go in the water. Now, I can't even have a picnic on the sand without being worried about a shark.

Plus, the teenager in Florida was only saved because his brother came to his rescue.

Which got me thinking, if I was ever attacked by a shark, who would come to my rescue?

Liza? She wouldn't want to get her hair wet. If I was ankle-deep in water, maybe.

Not that I'm so sure I'd even want to be rescued. Being bitten by a shark has got to be pretty traumatic. How do you recover from something like that? If I ever got bitten by a shark, I'd lose the few friends I have left. I get bitten by a mosquito and complain about it for weeks.

The worst is when I hear that these people are grateful to have survived. This makes me feel awful. As someone who has never felt grateful to be alive, it's hard to relate. I've often wondered what it would be like to feel that way, but I'm not sure having a limb bitten off by a shark is the way to find out.

I suppose it's not really that surprising that sharks are eating humans. All over the world people are making a living taking boatloads of tourists out to view the sharks and pouring buckets of blood over the side into the water to attract a feeding frenzy. The sharks are confused.

For me, shark-attacks are good news. It validates what I've suspected all along: stay in your apartment and you won't get eaten.

# You Can't Stay at My House

A British friend of mine is in New York, desperately looking to rent an apartment for six months. I told him I'd ask some friends, see what I could do. An hour later, I'd forgotten his request and told him I was planning to spend April out of New York. All of a sudden, he looked hopeful. 'Does that mean your apartment will be empty?'

Here was a situation that presented some problems. If I said yes, I was a shrew for not offering to let him stay in it. If I said no, I was a weasel for lying.

Telling someone who is looking for an apartment in Manhattan that you have one empty is like telling someone who's waiting for a bone-marrow transplant that you know of a donor but don't want to give them the phone number. It's not something they're likely to forget.

Being British, he began to apologize: 'I didn't mean to put you on the spot.' Now on top of not helping him, I was making him bear my guilt. How did that happen? I tried to explain I wasn't sure when I was leaving, or how long I'd be away for. That made it worse. I couldn't even produce a credible lie. But what could I tell him – the truth?

My apartment is tiny. I'm tiny. It's a tiny person's apartment. If you have long legs, it won't suit you. But no matter what I say, it doesn't matter, because the answer is always: 'I don't mind.' My home could be condemned as a carbon-monoxide-filled death trap, and the response would be: 'Does that mean it's empty?'

Why should I feel bad? I don't want someone's sweaty behind sitting on my toilet seat. Or their sticky ear pressed up against my phone. Plus everyone snoops and I know how it goes. First they would see the expensive Clarins massage oil and wonder: 'What does "relax" smell like?' The next thing you know, they're thinking: 'She won't notice if I use a little, just this once.' Soon there's hardly any left and they're using vegetable oil to top up the bottle.

And what about the personal things, especially in the medicine cabinet? It's one thing to take a quick peek, but what if someone were to leisurely examine each and every tube of gel and package of pills? Conclusions will be reached based on inconclusive evidence. There's no way to explain that I bought Preparation H after reading a beauty tip that suggested it gets rid of dark circles under the eyes. As far as the snooper's concerned, it's for haemorrhoids.

My apartment is a goldmine for any snooper. Here's why. I have an entire file devoted to ex-boyfriends and relationships gone wrong. Every time the relationship ended, I'd put everything he ever wrote me – every card, letter, e-mail – along with photographs, tapes of his messages on my answering machine – in a big manila envelope. Sometimes, if the relationship was particularly long or turbulent, two envelopes.

I have a dozen of these labelled: 'Mike', 'Mike 2', 'Punishment', etc. It doesn't get any better than that. Snooping + ex-boyfriend's angry e-mails = jackpot.

Everyone snoops and I can't be bothered to lock all my stuff away. Having someone stay in my home, I'd feel exposed. It would be as though I'd handed over my diary and said: 'Here, read whatever you want.' I can't take that risk. I have enough trouble holding on to friends as it is. I like my apartment friend-free.

# Brain Surgery

Ted Kennedy had brain surgery performed on him while he was wide awake.

Is there anything worse? I don't even want to be awake for a tooth cleaning. Apparently he had to remain conscious so they were sure not to do something wrong. If they were about to encroach on a part of the brain that controlled a function like speech, keeping him awake meant they could test it out. By repeatedly asking him questions such as who the President is or to identify objects in pictures and then as long as he got the correct answer, they knew they weren't messing up.

That seems like a lot of pressure. I know that when I'm on the spot, I tend to freeze. What if the surgeons asked me a question and I hesitated because I needed more time to come up with the answer? Would I be able to say, 'Wait, I'm thinking – give me a second'? I wouldn't want to be rushed. Or what if I got the wrong answer? Asking who the President is, that's easy. But what if they asked for the date? I never remember the date.

If the tumour is on the left part of my brain I'm in trouble. Left brain is the analytical and judgmental hemisphere. Right-brain questions are more my speed – more emotional. If they wanted to test it out they could ask what I was worried about. If I answered, 'Nothing', they'd know something was wrong.

Also, if I'm wide awake that means I'll be able to hear them talking. I can see it now, I'm lying there, fully conscious, unable to move, and the doctor says: 'Oops.'

Another thing. I have excellent hearing. I can hear when people are whispering. If someone said something nasty about me, like that they could tell I was high-maintenance, I'd instinctively turn my head to see who said it. That wouldn't bode well with a scalpel inside my skull.

Or let's say someone asks, 'Did anyone see last night's episode of *The Apprentice?*' I could be taping it and the surprise would be ruined. Then what would I have to look forward to?

Even worse, what if someone who is supposed to be paying attention asked the neurosurgeon, 'What did you do last night?' That would get my heart rate soaring. He's busy! Don't disturb him. It's bad enough having to listen to someone discuss how they spent their crazy night at the bar doing Jell-O shots but what if the surgeon mentions he's got a hangover? No-one wants a brain surgeon with the shakes.

I suppose the good news is that if I'm lying there wide awake, it means I'd have a captive audience. And I'd ask loads of questions. Starting with, 'How's it look?'

Chances are his response would be, 'Depressed.' Soon I'd be talking so much that the doctor would lose patience. Especially if I started asking questions like, 'Are you almost done?' I'd like to think he wouldn't intentionally remove anything he didn't have to – just to shut me up.

# The Dangers of Al-Fresco Therapy

There's a new trend in therapy. A therapist in New York has decided to run his practice outdoors. His patients spend the entire session with him in the park: walking and talking.

I can't think of anything less appealing. Why would someone choose to be outside in the oppressive sunshine when they can be safely inside in the womb-like safety of a darkened room?

But also, there are a multitude of other practical issues. For instance, if I'm going to spend fifty minutes walking that means I have to wear trainers. What if I'm not coming directly from home? I'll have to carry my gear around in a special bag. The way some people carry a bag for the gym, I'll have a bag for therapy? Trainers, pants, sun block and tissues.

Then when someone wonders why I'm not in better shape since I seem to be working out all the time I'll have to explain. 'I'm not going to the gym. I'm going to the shrink.'

Moreover, what about those of us who have allergies. Are we not supposed to have therapy in the spring? How will my shrink know if I'm crying or if my eyes are watering from the pollen?

I'll have to take an antihistamine before my appointment. But if I take one with speed I'll be power walking. And if I take one without, it will make me drowsy.

That's not very productive. If I'm spending $150, I don't want to be lying down on a park bench fighting to stay alert.

Where we meet is a potential disaster as well. Meeting at an office is easy. But that won't work because by the time we got to the park, we'd have to turn around and head back. So I assume this means we'd meet at a designated spot somewhere inside the park. I'm anxious just thinking of this. I'll end up in the wrong place.

I can hear it now. 'I thought you said to meet at the fountain?' Then we'll spend the next fifteen minutes discussing what it means that I never listen to directions.

Then there's the possibility of running into someone I know. That's terrifying. Do I stop to say hello and explain the situation? 'I'm in the middle of an emotional crisis so I can't chat. But nice to see you — you look great!'

Even worse than running into someone I know — running into someone I'm talking about. I could see obsessing about an ex who coincidently happens to be strolling through the park at that very moment with his new girlfriend. And of course this happens on a humid day when my hair is frizzy.

Triple session on the spot.

Research shows that a person is likely to feel more alienated and alone when they witness others whose lives have worked out. Was I the subject for this research? Walking through the park, this is a constant danger. It can actually cause harm. How? By subjecting a semi-depressed person to increased levels of inadequacy and plunging them into a full-blown depression.

On the other hand, witnessing those who are worse off can be euphoric.

I do see one advantage to an out-of-office therapy session. There's a far greater chance of going over the fifty minutes. Where are the clocks? When time is up, it's not like a therapist can kick you out of the park.

# My Clone

Recently, a clone was made of an anonymous Afghan hound; the first time a dog has ever been successfully cloned. The effort took South Korean researchers nearly three years, working seven days a week, 365 days a year.

As an only child, I've always wanted to have a sibling, but even better than having a sibling would be having a clone. How could I not adore her? We'd share everything. DNA would be the least of it. Finally someone would understand what it's like to be me.

But what if she didn't? Technically, she would be identical but she would have different parents and so chances are, she would be happy. That would be my worst nightmare. Someone exactly like me in every way, but happy.

I can hear it now: 'There's a party tonight – is your clone free?' Or, 'Your clone said the funniest thing the other day!' Or, 'Your clone is so much fun. Why can't you be more like your clone?'

Liza would call and say, 'I met a fantastic guy who's single – is your clone seeing anybody?'

My clone would get all the attention and I'd end up jealous. Which would make me look even worse. Who's jealous of their own clone? Although I guess if you're self-loathing it makes sense.

Liza has an identical twin sister, the closest thing to a clone. I asked her what it was like and she told me that for the most part it's great. 'Except everyone tells her she looks like Jennifer Aniston whereas I'm told I look like Bette Midler. How is that possible?'

The best thing would be to have a clone who is more messed up than I am. If my clone was worse off than me, that would be ideal. I'd get to feel sorry for myself *and* superior at the same time. I'd hang out with my loser clone all the time and it would remind me how lucky I am. It's not going to happen though.

## Losing My Life

Recently I discovered there's only one thing that I am attached to. A few weeks ago, I had a scare and I realized how much it mattered. My plane had just landed at JFK. An announcement was made that it would be a while before we could disembark so I turned my phone on and made some calls. As I exited the aircraft and walked to immigration, I was still talking, following the crowd.

Standing there, breaking the rules by chatting illegally on my cell phone in the immigration line, I wondered why my shoulder wasn't aching the way it usually does. Then it hit me: I had left my laptop on the plane. Panic. I hung up and instantly sprinted back to the gate.

Anyone who's ever landed at JFK international airport knows that the distance between immigration and the gate where the plane arrives is longer than the flight itself. And I was soaring. Like a gazelle in my natural habitat. Only without the gracefulness.

It was late at night and the terminal was completely empty. I saw a security officer and I shouted: 'You have to help me! Please . . . !' I was too winded to get the rest of the sentence out.

When I caught up to him I explained what happened. He told me he would accompany me to the plane. 'Please . . . walk faster . . .' I said, the entire way.

All of my writing is on my computer. I rarely print anything out and I don't back things up. 'You don't understand,' I said, 'That computer is my life.' Julius kept nodding. He looked like he wished he hadn't stopped.

We reached the plane and the cleaning crew was on board; they wouldn't let me enter. Julius explained and when he was done I repeated everything he said as if I was translating. Even though he had spoken English. 'It's an Apple laptop,' I said. 'It's my life.'

I couldn't stop saying this.

Julius and I waited as they searched. Five minutes later when they returned they were empty handed. It was stolen, they said.

This was not good. We began power-walking back towards immigration. He was on his phone trying to reach his supervisor but now there was no reception. While he was waiting for a signal, I was manically instructing him to try again. 'They have to make an announcement . . .' I said, as I envisaged someone hailing a cab outside the terminal with my computer under their arm. 'Tell them to make an announcement.'

By the time we got back to the immigration line, we'd been gone nearly half an hour. The line hadn't moved. I'm breathless, sweating. 'Has anyone taken my computer??' I shout. 'Has anyone seen it?'

People are shaking their heads. Now it's become a situation.

Julius tells me I have to get in the line to go through passport control but I can't move. He heads off to speak to someone on the other side. People are saying nice things to me but I can't focus. All I can think is that if my laptop is gone, I have no idea what I'll do. Just then as he returns, I see that under his arm is my computer. I run towards him, and give him a hug. He doesn't hug me back. But I can't let go.

At that moment, I felt more grateful then I've felt in a long time. So grateful, I start crying. I let go of Julius and move to the end of the line and there is a woman with an infant in front of me. 'It's like losing a child,' I say.

She gives me a disapproving look. 'It's not the same.'

# Driving

R arely have I been called 'Ma'am' where an argument doesn't follow. Generally it's used to signify impatience by a stranger at the other end of the phone who has no intention of being helpful. No matter how angry or frustrated I become, the voice remains monotone. Attaching a mechanical 'Ma'am' gives the illusion of civility.

I had to drive while I was in LA recently. Every time I sat at the wheel I could feel myself getting older. Apart from the glaring sun and the monstrous Hummers, there are too many signs. People shouldn't be expected to read and drive and change lanes and find exits on highways that have ten lanes. It reminded me I'm not made for this world.

Naturally, I had an accident. I've always thought the word 'accident' is a bit understated for what can happen to someone in a car. The way 'disgruntled' seems to be the wrong word to describe postal workers who go on a rampage and shoot up the post office. I think of an accident as doing something stupid, like spilling coffee on the rug. It does not involve potential death or disfigurement.

But this time, accident was the perfect word. I was on my way to a meeting and the instructions directed me to an underground car park. I followed the sign and drove down the steep ramp. Midway I noticed another sign that said: 'Residents only'. Unable to turn around, when I got to the barrier at the bottom I pressed the intercom button for the attendant. 'May I help you, Ma'am?'

'Yes,' I replied. 'Can you raise the barrier so I can turn around, I'm in the wrong place.' The voice responded: 'Sorry, Ma'am, I can't help you without ID.' I asked how I was supposed to turn around. Silence.

Eventually I accepted that she wasn't going to raise the barrier. I would have to drive through it or reverse up the hill around a blind bend onto the road and risk getting killed.

I'm not a risk-taker. And backing up a steep hill onto a busy road? It doesn't get much riskier than that. I might as well have been wearing a blindfold too because I couldn't see anything – right up to the point when I hit the concrete barrier. Immediately, I switched from reverse to drive and stepped on the gas, but the car wouldn't move. I could hear screeching and scraping sounds, but as I wasn't hurt, how bad could it be? Turns out, pretty bad. So bad, I couldn't open the door.

There's a reason I don't drive. It's too much responsibility. But also, there are people who belong at the wheel – confident that they will get to their destination safely. And then there are people like me. People who belong in a taxi. The only 'Ma'am' I don't mind is when it follows: 'Where to?'

# A Baby in Brooklyn

My friend Audrey was very excited. Her sister-in-law was due to have a baby any minute and she was waiting for the call so she could rush over and be there for the birth. Then she mentioned they're having the baby at home.

At home? 'Yeah, why?' she asked. 'Because . . .' I paused. 'Isn't that dirty?'

She made a face. So I had to explain. I didn't mean dirty as in gross or erotic; I meant dirty as in dusty. I thought for sure when I cleared that up it would make things better. It didn't.

'That's the first thing you think of?' she winced. 'The apartment is dusty?'

Who wouldn't think that? They live in Brooklyn. If you leave your window open for half a day, the sill looks like it's been painted black. Plus, a woman who is nine months pregnant? Vacuuming would be the last thing on her mind. It seemed like a perfectly rational thought to be concerned that a New York apartment isn't the most sanitary place to give birth.

Audrey was baffled, though, and I couldn't understand why. She explained: 'Here I am thinking about the miracle of life and how joyful it will be to see a baby come into the world, and all you can ask me is if it's dusty.'

I'm not saying there's anything wrong with it. But it doesn't seem right to give birth to a child in the same room where you take off your socks. Also, let's say it's a long labour and the dad-to-be goes to the kitchen to make a sandwich. All of a sudden, the baby's coming. He runs to the bedroom to help deliver, and next thing you know, there's pastrami on the baby.

Audrey interrupted me. 'Pastrami on the baby? What are you talking about?' Okay, mustard. He gets mustard on the newborn. And what about salmonella? Or E. coli?

'They wear gloves,' she said. He's making a sandwich in the same gloves he wears to deliver his child? It doesn't seem right. And I think there's something sad about giving birth to a child with a broom in the corner. That's what the baby will see when it comes into the world? She told me they didn't have a broom in the corner.

It's hard enough to come into the world. The poor baby has no idea what it's in for. At least give it something to look forward to: going home. But if you're born at home, where do you go from there? The hall?

If there's ever a reason to get out of the house, having a baby would be it. That, or a fire. Which is another thing to consider. What if you're having a baby at home and the toaster explodes? Choose between the baby or the house burning down? You don't have to worry about that if you're at the hospital.

I know the most important thing is that the baby is healthy. I've been to a hospital recently and noticed the antibacterial hand soap they use is a lot stronger than anything you can buy at the grocery store. And then hospitals have this other thing that comes in handy in case something goes wrong. Doctors. Why would someone pass that up?

# Taking a Risk

I've been spending a lot of time thinking about risk. I know that for most people it's a positive thing. Taking a risk equals feeling alive. Every day there is somebody jumping out of a plane, quitting their job, trying something new. But what defines risk? I take a risk every time I pick up the phone. My caller ID doesn't always work and sometimes, when it says 'unknown caller', I answer it anyway.

Sex is a big risk. My friend Emily is dating someone new and had sex with him on the second date. Afterwards she thought maybe she'd slept with him too soon. Now she's worried he's only interested in her because of sex. It was a risk. But what was the alternative? If she'd held off, she'd only wonder if he was interested because she played hard to get. You can't win. I pointed out that she wasn't worrying about the right risk anyway: she should be worrying about STDs. If you're going to worry about risk, it may as well be one that involves a disease.

The risks I take seem to be mainly by accident. I'll kiss someone and find out afterwards they've been feeling under the weather. Or I'll forget to wash my hands after taking the subway, then eat a piece of pineapple with my fingers.

I usually find out that I've taken a risk after the event. That's not 'good risky', the kind someone finds attractive. It's 'worrying risky'. The kind nobody wants to hear about.

Here's the latest risk I've uncovered. A doctor friend in London found out I was using tap water to rinse my hard contact lenses. 'You're risking your vision!' he exclaimed. That got my attention. For twenty years I've been risking going blind? He explained that there is an amoeba in tap water that infects the eye and can cause acanthamoeba keratitis, which can lead to loss of vision.

But on the bottle of my contact-lens solution, it categorically states in bold black letters: 'Rinse lenses with fresh tap water'. I called the manufacturer's global headquarters. Customer service asked me what it was in reference to. I told her: 'Going blind.'

She put me on hold.

I called my eye doctor in New York instead. In a calm voice he said acanthamoeba keratitis is a devastating condition but only affects 1 in 500,000 people. There are ten million people in New York. That sounds more like 1 in 20 to me. I could be the one.

Hearing the panic in my voice, he assured me that I had nothing to worry about, and that Manhattan's tap water was protected: it was well-water that might be a problem. I asked: 'What if I'm swimming in a lake?' He told me not to open my eyes. Then he told me to avoid Jacuzzis. Jacuzzis? Why would someone open their eyes underwater in a hot tub? I didn't want to know.

Besides, I wouldn't go in a hot tub if you paid me. It's a cesspool of germs.

He told me he had to go. 'Have a good day,' he said. I'm not sure what to do. Why did vision have to become so complicated? Now I feel I'm at risk of being in danger as soon as I wake up and open my eyes. But that's never something people will consider to be exciting, is it? You'll never hear someone say: 'That girl is a real risk-taker. Every day, she gets out of bed.'

# four

THEM AND US: A VIEW FROM THE BASEMENT

# Nothing vs No Plans

Why can't doing nothing be seen as something attractive? It's very difficult, impossible really, to live in New York or London, or any city for that matter, and feel it's okay to do nothing. You're not allowed. If you're home on a Friday night with nothing to do, you have to qualify it. The excuses kick in.

'I *wanted* a night to myself.' Or, 'It's the end of a long week and I'm exhausted.'

People say affirmations are very important. Here are some of my favourite delusional ones: 1. I'm okay being alone. 2. I like my body just the way it is. 3. I don't care about New Year's Eve – I prefer staying home.

I was talking to my upbeat and very optimistic friend, Emily. Emily and I have nothing in common. Emily doesn't dwell, sees only the good in everyone, compartmentalizes, and never gets angry. It's amazing we have anything to talk about.

'What are you doing this weekend?' she asked.

'Nothing. What about you?'

She paused. 'No plans.'

Just then I discovered something very important. There is a world of difference between having nothing to do and having no plans.

People with 'nothing to do' have no options. People with 'no plans' have the world as their oyster – they just haven't decided yet what they're going to do. They might go skiing in the Alps. They might buy a Ferrari. They could do everything, or nothing at all. Who knows. It all lies ahead.

But for people with nothing to do, nothing lies ahead. They are self-pitying, poor and lazy. They feel ripped off by other people's plans and doomed to spend the rest of their life alone with their empty calendar.

People with no plans are confident go-getters with self-esteem, a take-it-or-leave-it attitude and they see life as one big adventure just waiting to happen. People with nothing to do see life as one big adventure that someone else is having and they weren't invited.

People with no plans are optimistic. People with nothing to do are pathetic.

From now on, if someone asks, 'What are you doing?' I've decided to say, 'I'm free.' This puts all the responsibility on the other person. It takes the pressure off and I'm being very up front about it. I'm saying: you are now responsible for my happiness. Do something about it. Make me happy.

I've found men don't respond well to this.

## The Pressure of a Sunny Saturday

When it's a sunny day, there's a lot of pressure to be outside. Everyone has to be outside. Staying inside is out of the question.

Saturdays are the worst. Brunch is mandatory. If you're not outside having brunch on a sunny Saturday, there's something wrong with you. Tables are set up on the sidewalk. You have to walk everywhere – even if it's miles away – because it's nice out. The nail salons are teeming with people and open-toed sandals are requisite footwear. By five o'clock the sun has gone down and it's 10 degrees but it doesn't matter; it's better to be walking around shivering in a little sun dress with toes that have turned blue – than to go inside; especially if it's still light out.

One Saturday my phone rang. It was Emily.

'You're not outside?!' She sounded so outraged, I felt bad for her. 'It's a beautiful day!'

Emily was calling from the flea market. Or was it the flower market.

Or was it the farmers market? If it's Saturday and it's sunny, everyone has to be at a market. Or a marina. Or a festival in the park. Or get ready: it's Bastille Day!

Unfortunately, going outside isn't a passing trend. So as this craze shows no signs of slowing down, I succumbed and went out. Who knows – I might miss the last bit of sunshine ever.

Gazing into the window of Citerella, an upscale food market, there was a woman in open-toed sandals holding hands with her hot boyfriend. She carried the wine, he carried the baguettes.

'Honey, we should get some of that really *fresh* buffalo mozzarella, don't you think?'

And therein lies the tragedy of a sunny Saturday. It's all about couples grocery shopping for the dinner party they're either having or going to that evening. Who needs to know about that when I'm buying my apple for one? The only thing that cheered me up was knowing that soon it would be Sunday and everyone would be inside, bedridden with a miserable hangover.

I returned to my apartment bolstered by this perspicacity. Just as the sun was disappearing.

# The Most Depressing Day of the Year

January 24th has officially been declared the most depressing day of the year. A doctor at Cardiff University has determined that misery is expected to peak on this day and devised a formula that proves it. Weather + Debt x Motivational levels + Time since failed attempt to quit smoking x Days passed since Christmas = January 24th. Sounds good to me. There is nothing more appealing than knowing that on January 24th, everyone will have fallen off the wagon, failed to be anything other than broke, unmotivated and down in the dumps.

Could I have asked for a better day to have as my birthday?

For me, January 24th has always been a day of despair. So I can't say I was surprised to learn it's official. I'm just amazed it took so long to recognize something I've known all my life.

But if I were to go by this doctor's theory, the 24th should be a great day for me since I enjoy bleak weather, didn't indulge in the post-holiday sales because I already considered my financial situation dire, have made no attempt to quit smoking, and the only day that depresses me more than January 24th is Christmas, so the more distance from the holiday the better.

It should be great but it isn't. Why? Because it's still my birthday. Birthdays are for people like Kate Moss. People who wear yellow hot pants and have a hot rock band perform live at their home. If you're Kate Moss and your birthday is in the middle of January, nobody cares. Friends will purchase a snowplough to get there. If you're not Kate Moss and your birthday is in the middle of January, nobody shows. The excuses range from the wind was blowing and there were no taxis to the always reliable: 'I have the flu.'

I was curious to know who else was born on this day of dejection. John Belushi. Sharon Tate. Good company. Will Smith, Edith Wharton. Who knew those two would have a connection? And Mary Lou Retton, the perky gymnast who won all around in women's gymnastics at the Olympics. That was in 1984. Chances are she's sufficiently depressed by now.

But just as all the misery peaks in the UK, I tend to be in the US. Where January 24th is National Peanut Butter Day.

# I Know You Don't Like Me

Life would be so much easier if people just said, 'I don't like you.'
Think of all the time it would save. No more lying to get out of
relationships. No more trying to figure out what you might have done
to make someone distant or wondering if you're being paranoid. They
wouldn't even have to offer a reason. 'I don't like you' is all I need.
Knowing where they stood would be enough.

I met someone at a party and instinctively knew that she didn't like
me. I could tell. Even before we were introduced she had been standing
nearby and I saw her give me a look. It was a look one woman gives
another woman which conveys a mild and silent revulsion; invisible to
everyone else in the room.

Usually there is no justification for this look so it seems irrational.
And when you mention it to someone – a mutual friend – the friend
will invariably tell you that you're imagining it. Especially if the person
who gave you the look is more popular. Which, for me, is always the
case. Then I regret having said anything at all. I'll think to myself, she
appears to be a warm person. To others.

When you get the feeling that a likeable person doesn't like you, it's
best to keep it to yourself. No good can come from mentioning it.

If you ask a friend for confirmation, you're not going to get it. And
then if you do, it's even worse. There will be a fleeting satisfaction knowing you were right which will immediately give way to a desperate need
for more information. And this won't make anything better. Plus, now
you know that the person who doesn't like you has talked about not
liking you to other people. It's not good.

Of course everyone reacts differently. Some people respond to not
being liked by trying to win people over. I don't have that kind of energy.
When I get the feeling someone doesn't like me, I'm relieved. It's one
less person I have to worry about alienating.

The problem is, no-one ever comes out and says it directly, which
leads me to spend hours wondering if my instincts are on target or not

and trying to figure it out. This is time I could spend obsessing about other more pressing matters. Like why someone hasn't called.

When it comes to being disliked on first sight, gender plays a part as well. If a man dislikes me on sight, I assume it's because he's not into me and who cares. With a woman, it's more serious. It could be any number of things. The only time being disliked doesn't play out along gender lines is when I'm in a restaurant. Then there's no confusion. Both waiters and waitresses hate me.

I understand not wanting to hurt someone's feelings but I've always felt straightforward rejection is far more considerate. I've been rejected plenty of times and the more direct the better. Especially when it comes to relationships. There's something missing? Fine with me. You can't see a future with me? No problem. I can't see a future with me either.

I'd much rather hear that someone doesn't like me because chances are, I don't like him either and at least then we have our mutual dislike in common.

Then again, 'I don't like you' is one thing. That I can handle. 'I don't like you any more' is another story. Then it gets messy. This means that at one point I was liked but now – not so much. What could have happened? There's only one explanation. They got to know me.

# Being Mysterious

Liza and I were talking on the phone. 'You know what I don't have?' she said. 'An air of mystery.' It's true. She tells me, and everyone else in her life, everything. I've got so used to it that I now feel outraged if something happens and she forgets to mention it. 'You switched shampoo?' I'll say. 'You never told me that.' I'll assume others have known all along, and feel left out.

I asked when her air of mystery was to begin. 'Tomorrow,' she said. But as soon as she said that, I realized it would be an uphill battle. Announcing when you're going to start being mysterious is not very mysterious.

What makes someone elusive? It helps to be French. French women have an enigma gene that enables them to be silent and chic, as opposed to lifeless and dull. Catherine Deneuve is elusive. But if she were from Florida, she'd be just another blonde who dressed well and didn't have much to say.

There is a lot to be said for not being so available. It makes people interested. I've tried this, but it didn't work out. Everyone just seemed relieved. Either that or they thought something was wrong. If I'm ever quiet, people will wonder what's happened. They ask: 'Are you okay?' Then I'll say: 'No, not really...' and assume they are interested in why. They're not.

If I've ever had an air of mystery, it's by default. Like when I avoid someone. That's when I've had men tell me that I'm mysterious. But it never works the other way around. If I think someone is avoiding me, it would never occur to me that they're being mysterious. I assume I've done something wrong. So I'll leave another message asking them to call me back, and wonder why they're being evasive.

The difference between being elusive and evasive is something I'm not clear on, either. Next time there's a question I don't want to answer, I'll chalk it up to being elusive. Let's see if it makes me more attractive.

Men like mysterious women. They're usually not talkers. A woman of mystery will show up late for dinner and simply apologize, no explanation. She'll sit for hours on end when the football is on, mysteriously, not making a sound. If something is wrong, she'll keep it to herself. I, on the other hand, show up late and rattle off a monologue on the traffic, my blisters, the weather and its effects on hair and mood. Essentially a rundown of my entire psychological wellbeing. Far less appealing.

Some people are so desperate to have mystery in their life, they'll latch onto anything. For instance, I had a friend who dated a man with a beard. 'It's so mysterious,' she said. Really? I see a man with a beard and presume all he's hiding is not having a chin.

Maybe I've never cared about mystery because I'm too impatient. Growing up, reading Nancy Drew detective novels, I'd always skip to the end and find out how she solved the case. If only I could do that in life. Skip to the end and see how it all works out.

I checked with Liza to see how life as a mysterious woman was going. Not well. She'd gone on a date, didn't tell anyone, and then, when nobody called to see how it went, she was depressed. 'Now I feel like nobody cares about me,' she said. I welcomed her to my world.

# Life on Earth

Stephen Hawking announced recently that the survival of the human race depends on its ability to spread out into space and find a new home there. Life on Earth, he says, is at risk of being wiped out by a disaster such as nuclear war, global warming, a genetically engineered virus, or other dangers we haven't even thought about.

I don't really see the problem. Does the human race really have to survive?

Maybe I just haven't been hanging out with the right people. I have yet to meet someone who reminds me how valuable and crucial our life form is. The other day I was walking along and I saw a man on the street in front of Jamba Juice dressed as a giant banana. It was 100 degrees out and he must have been roasting. Pedestrians just passed him by – nobody paid much attention. At one point someone accidentally knocked into him, which prompted the banana man to call out: 'Excuse you!' This is what humans have come to.

Hawking doesn't care. He predicted that human beings could have a permanent base on the moon in twenty years and a colony on Mars in the next forty. 'We won't find anywhere as nice as Earth unless we go to another star system,' he said.

Anyone who's ever been stuck on the Tube in August will know that if Earth is as good as it gets, relocating isn't all that appealing. Life on Earth is hard enough. I'm not sure how well I'd do without the oxygen and gravity.

And finding a home elsewhere in the universe? Sounds like a lot of work. If I can't get someone to pick me up at the airport, what are the chances I'm going to find someone who's willing to help me move to another solar system?

I can see it now. It will feel just like the holidays, where everyone is making plans to go away and I'm left behind. Only instead of it being for two weeks, it will be for ever. The entire planet will be abuzz getting ready to colonize Mars, and I'll be home, alone, watching the exodus on CNN. Feeling relieved that I don't have to shower and change out of my pyjamas.

Eventually Liza will call, sounding really excited because the weather on Mars is great for her hair. Mars has low humidity. All right, so there's water and a bit of atmosphere, but isn't it also incredibly dusty? My eyes itch just thinking about it.

She'll ask if I want to go shopping for biosphere suits, and I'll say: 'Nah, I feel fat today. I'll go tomorrow.' But then I'll put it off until eventually all the moon boots in my size are sold out. That will give me an excuse that people will accept: I can't go to Mars – I don't have the outfit.

The more I think about it, the more I realize how little it matters to me if the human race disappears. In fact, a meteor can't hit soon enough. Scientists have been predicting something catastrophic for a while now. What's the hold-up? Aren't we due?

Geologists just discovered a crater in Antarctica that may have been caused by a meteor that eliminated more than 90 per cent of the species on Earth 250 million years ago. Obviously, some people made it. So

really, the worrying part would be if the entire species was wiped out –
except for me. Or worse: just me and a German supermodel. That
would be fun.

# You're Nobody Till Somebody Loves You

Y ou're nobody till somebody loves you. This lyric's been around for
ages. It seems that reminding us single people that we may as well
kill ourselves is not only timeless, it's popular.

There must be loads of nobodies out there. But how do we find each
other? It's not like we go around advertising: 'Hey, I'm nobody. Want to
talk to me?'

Being nobody means you don't have a lot of options. It's hard to
meet other nobodies because we never go out. If I do have a soul mate,
which seems unlikely, he's someone I'll see on the subway and just as
I'm about to talk to him, he'll get off. Or we'll be seated next to each
other on a plane and become close just as the pilot says there's engine
trouble, and we crash.

My soul mate is probably home right now, thinking he should go
out but not doing anything about it. Either that or he's marrying
someone else because I changed my mind at the last minute and never
went to the dinner party I promised I'd go to. And that night he sat
next to my replacement, who he found only mildly attractive, but
thought, 'I'm never going to meet anyone better than this,' and asked
for her number. Because my soul mate, like me, makes horrible
mistakes.

Speaking of mistakes, I went to an online dating site to see what all
the fuss was about. But I quickly discovered you have to invest a lot of
energy filling out forms. There were a lot of steps. I told Liza I don't have
time for that. 'What do you mean? You have nothing but time,' she said.

There was a questionnaire for what sort of man I was looking for, and two questions into it I lost interest.

What I wanted to write was: 'The type of guy I'm looking for is someone who would never be reading this form.'

There was also a section where I had to write a profile, and when it asked how I would describe myself, I clicked off. I could have put 'impatient and lazy' or 'nobody', but I didn't see that winning me a lot of dates.

Liza has been online dating for a while and has become very discouraged. As a sexy, slim thirty-eight-year-old woman, she gets, on average, one or two responses per week.

She was curious to see what it's like for the men out there and filled out the form pretending to be a guy. Only, she didn't want to attract too much attention so she checked 'unemployed', 'bald' and 'extra extra large'. She got ten responses within twenty-four hours. What does this show? Aside from the fact that women are much less choosy, it reveals that there are a lot more women desperate to be somebody.

But I'm not even sure I want to be somebody. Being nobody is much easier. You don't have to do anything to make yourself appealing because expectations are pretty low. Being somebody is work. Being nobody, there's nothing to lose. Of course, if another nobody were to come along, I would be thrilled. But then I would wonder: do two nobodies make a somebody? If so, which one of us gets to be it?

And what happens if somebody loves you and leaves you? Does that mean you were nobody, then somebody, then nobody again?

The only thing worse than being nobody is being a former somebody.

# Devotion You Can't Buy

I've always had a problem with talking about what I do. When I did nothing, it was a lot easier. I'd make it sound like there was a big career on the horizon: 'I'm working on something . . .' People responded nicely to that because even though I could be a loser, I could also be the next big thing. You never know.

But even when I told the truth and said I had nothing going on, people thought I was just being modest. I would say: 'I'm a writer,' and if someone asked what I'd written, I'd reply: 'I can't talk about it.' Then I was mysterious.

Now that I've done stuff, I resist going into detail because I know that when someone asks: 'What do you do?' what they're really asking is: 'Are you someone that I want to talk to?'

And it's not that I don't want to sell myself, it's that I'm too lazy. There's a second where I have to decide if I'm going to gear up and trawl through my history, and I think: 'What's the point? We're all going to die soon – what does it really matter?'

I've been around a lot of couples lately. And you know what I've noticed? Being in a relationship is like having a full-time publicist. You don't have to feel embarrassed about listing your accomplishments, because the other person will do it for you. It's devotion you can't buy.

I was at a wedding, seated at a table of couples. When the 'What do you do?' question was raised, one of them would answer demurely and then their spouse would jump in and list their accomplishments. For instance, one man announced he was a director and his wife interrupted: 'His short film just won at Cannes. It was brilliant.' It's always brilliant.

If I were in a relationship, I would love to sit back while my boyfriend sold me. The problem is, who's going to sell me to my boyfriend? I'm done selling. I need to be in a relationship where the 'getting to know you' phase is out of the way. I wish I could hand my future husband a CV of my life. Here. It's all on paper, my childhood, the break-ups, the best friends who dumped me – all of it.

Of course, a lot of men would take one look at the CV and run a mile. But that's a plus. Think of all the dinners it would save me having to sit through.

So let's say I find someone, we fall in love, become a couple, and he knows me well enough to be able to sell me. Then what? I'd have to reciprocate? Because the only thing less appealing than selling myself is having to sell someone else.

I suppose the best-case scenario would be to end up with someone who doesn't sell me, so that I don't have to sell him in return. If anyone asks either of us what we do, we simultaneously get up and go home.

# A Group of Friends

I've always wanted a big group of friends. We'd meet up for brunch and discuss every detail of each other's lives. When something happened to me, everyone would care.

Let's say I found out I had lupus. We'd go out, have coffee, discuss my treatment, and then what? Having a disease is bad enough, but having nobody to repeat the story to? That's really sad.

The friends that I do have don't know each other. I meet them one to one, and whenever I've tried to make it a group, no-one gets along. Or, even worse, they get along and end up becoming better friends with each other.

A while ago I introduced Emily to Audrey, and it had horrible consequences.

They hit if off and exchanged e-mails. Now they're the best of friends and hang out all the time, except they don't invite me. Instead of creating a group, I created a void.

When I asked Emily what she was doing for Thanksgiving, she hesitated. 'I'm going to Audrey's country house.' How did that happen? I've known Audrey for years and I've never once been invited to her country house. Of course, I wouldn't go if she asked. But still.

Recently I was invited to a baby shower. But to be a part of this group you need to have a husband and a baby, or at least the ability to pretend you're interested in them. There is no better way to feel like an outcast than to bring up the war in Iraq at a baby shower.

'You need to hang out with my gay friends,' Liza said. 'All you need to be a part of their group is some glitter.'

I gave it a shot. At Simon's birthday party there were so many gay men that I could have shown up naked and nobody would have noticed. But by ten-thirty, just as things were getting started, I got a piece of glitter in my eye. Rather than tough it out and go blind, I went home.

I tried to think of what the requirements would be to be included in my group. I'm not sure what they would be – I only know what they wouldn't be.

No husband, no children, no pets. No drink or dancing or partying. You shouldn't want to go anywhere or do anything, and there'd be no pressure to show up. All you'd need to be a part of my group is a desire to be alone.

There's a group of ladies who live in my building and get together every Thursday night to play bridge. The average age is seventy-two. I wouldn't mind that. Think of the diseases we could discuss. I'd have evenings of arthritis and melanomas and sciatica and dementia. I'd fit right in. But then they'd probably want to stay up late to finish the game. And I'd upset the balance by leaving. I'd tell them I had an early doctor's appointment, but they'd roll their eyes and tell me they'd heard that one before.

# Other People's Children and My Fake Child

A friend of mine had a fortieth birthday party, and five minutes after I arrived, a drunken boyfriend of someone I don't know approached me and inquired if I was single. 'Why?' I asked. He shrugged. Wasn't there anything else to talk about? He said, 'Sure,' and thought for a second. Then he asked if I wanted a drink. When I said I didn't drink he wanted to know why. I guess those were the conversation choices: why I'm single and why I don't drink.

The next person I spoke to was someone I hadn't seen in a few years. He asked if I was still single. Yes, I said. Still. Can you believe it? He seemed upset. So much more upset than I've ever been, and I've been pretty upset.

The look on his face conveyed absolute grief. Like he was in mourning for me. Which made sense. Because in his eyes, I would have been better off dead than still single.

At the same party, I met a couple who had just had their second baby. 'I'm not the type of person who tells stories about their adorable children,' she said. Which was my cue to say: 'Go on.'

So she did. When you don't have children and someone you've never met tells a story about something their child did, it's like listening to an in-depth story of the bus driver learning to swim. And having a visual doesn't help. It's still a stranger. This woman showed me a picture of her baby on her iPhone, and as she was doing this asked: 'You want to see what she looks like?' Now I'm on the spot. If I say no, I'm the rude one. How did that happen?

The problem is, most stories about children are never as funny as parents think, and unless you too have a child, it becomes a one-sided conversation. There's no way I can join in unless I try to remember what I was like when I was that age. But whenever I add, 'I used to do the same thing,' they look horrified. Because suddenly there's the possibility their child will end up like me.

I have a friend who has a four-year-old and I've never once heard him tell a story about his son. When I asked him why, he said: 'He hasn't done anything that interesting yet.'

If you're going to tell a story about your child, brevity is key.

I don't need background on the baby-sitter, the digestive system, and the social history of the other children involved. If I'm barely interested in the details of your child, who I don't know and will never meet, what makes you think I'd care about all the extras?

When the story was over, the woman asked if I was interested in having children. Other than yes, there was no good answer to that question. I said I wasn't sure. 'Well,' she said sympathetically, 'you must be so busy with work.'

No, I said, not really. Why is it that there has to be a career that is preventing me from having a child? As though that must fill the tremendous void I have in my life, being childless and single. Maybe I just don't want kids. Isn't that enough?

A few days later I was having coffee, and on the floor of the cafe there was a photo. I picked it up and it was of a baby. I stuck it in my wallet and am carrying it around, so from now on I will be presenting this child to strangers with stories of her first cough and spit. Her name changes from day to day and she loves Mexican food. My life has been filled with meaning since I've had my fake child.


# My Fake Family

The health-food market is packed. I go in, see that it would take thirty minutes to pay for my wheat-free cookie — and just as I'm about to leave, a very tall Rastafarian man who works there spots me and a huge smile crosses his face. 'Hey, I haven't seen you in ages!' he says.

I'm touched that he's so happy to see me. I used to go to this store a lot but haven't been there for years. It's rare that I encounter people that happy to see me. I'll take it where I can get it.

He's ushering me towards an empty register – where there is no line – and as I follow he asks: 'How are the kids?'

Now the happy greeting makes sense. It was meant for someone else.

'Good,' I reply without even the slightest hesitation. 'Going anywhere for the holidays?' he asks. All my life, this has been a question I hate.

Not just the part about going anywhere, because the answer is always no, but the entire line of questioning depresses me. But since I'm someone else I reply: 'Vermont.' Then I pause. 'With the in-laws.'

I tell him it's beautiful. 'And Jack loves the snow,' I add. It feels like I'm shoplifting. Only I'm stealing a life. How much can I get away with?

Apparently, a lot. 'Should I put this on your account?' he asks. But before I have a chance to say yes, he's laughing because he's kidding. Who puts a single cookie on a charge account? I hand over a note and the chat continues. 'How old is Jack now?' I'm guessing Jack is my son. I'd intended him to be my husband, but it doesn't matter. Jack can be my son.

It's not worth correcting him and I remind myself that I'm lying, so I don't need to be accurate. I decide Jack is three. That's a good age. A friend of mine has a three-year-old who I met the other day and they're cute.

I'm done with my purchase, free to go, but I can't. I have his undivided attention – why waste it? With the opportunity to continue, I do. 'Someone mentioned Ayurvedic medicine is good for rosacea,' I say. 'What do you know about this?'

He comes out from behind the counter for a more in-depth discussion. Immediately, I'm regretting my question. There is a lengthy explanation about the pros and cons. Then, for some reason, I bring up the Macy's Thanksgiving Day parade and last year's M&M's balloon catastrophe. A gust of wind caused a giant balloon to hit a lamppost, knocking the light down, which fell on a disabled woman. So this year they've been training the balloon handlers like they're being sent to Afghanistan.

Parades in general, I say, are annoying. He looks surprised. Then his entire mood changes. 'What do you dislike about a parade?' He sounds offended.

Several things. They're all the same, marching bands are too loud, they tie up traffic – you can't cross the street and you can never see anything unless you're in the front row, and if you want to be in the front row, you have to get there a day in advance. His arms are crossed and I can tell I'm upsetting him.

'I think parades are fun,' he says. 'What do you have against fun?'

That's a complicated question. I know it will take too long to answer it honestly and suddenly it hits me: even in my make-believe happy life, I get on people's nerves.


# Now's the Time

I told Liza that I was thinking of going back to the gym. Her response? 'Well, now's the time.' What does that mean? That if I don't go now, it'll be too late; I'll be out of shape and unlovable for the rest of my life?

If I'd said I was thinking of having a child, there's an obvious time limit. But going to the gym? After she said it, I realized I don't really want to rejoin the gym, because I'm never in one place long enough to make it worthwhile. So really, now is the time to be sedentary.

When I graduated from high school, everyone said that was the time to go backpacking around Europe. Travel the world. College was about to start, which meant four years of hard work – it was a free pass to let loose and to concentrate on having fun. I didn't know what to do with myself. I took a summer job reading film scripts.

After college, it was 'the time' again. Your early twenties are all about making mistakes and trying new things. It was implicit that soon it would be time to become a grown-up with a career and responsibilities, so now was the time to sample everything that was out there. The only thing I remember about my twenties is counting the seconds until they were over.

Early thirties — that was the time for dating. Not being anywhere close to getting married or having children, it was okay to go a little crazy. There was an expectation that I could date as many men as possible, because soon I would find The One and settle down. That time came and went.

Late thirties was the time to make a change in my life. I thought about what else I could do and began to get very depressed. Peace Corps? Too many bugs. Greenpeace? Can't be on a boat. My skills were limited. There were loads of exciting options — but apparently it was the time to put them off.

I'd love to get to a point where now is never the time. But that doesn't seem likely to happen. Seizing the moment is a constant headache that never goes away. For instance, just as you're about to turn forty, it becomes the time to take chances — go out with younger men, get cosmetic work done, have a stupid fling.

Then in your fifties it's the time to see the Great Wall of China, the Taj Mahal and the Grand Canyon, wear a visor and visit all the places you've never seen while you can still get around.

Your sixties will be the time to get all the replacement surgery in — the hip and knee — while you can recover properly and still have a few years left to reap the rewards.

The time to be hedonistic is your seventies and eighties. If I wanted to eat an entire chocolate cake at eighty-four, someone would say: 'Well, now's the time.' Things like cholesterol no longer matter. It's the time to eat and drink and smoke whatever you want because soon you'll be dead, so who cares?

I'm sure now's the time in my life to be doing a lot of things I'm not doing. I feel bad about that. But even worse is knowing that when I'm eighty-four I'll look back on where I am now and think: those were the days.

# five

PERSONALITY DEFECTS

# She's so Sensitive

Being sensitive used to be a good thing. A sensitive person was kind, compassionate, insightful, considerate of people's feelings. But recently, someone described me as sensitive and I'm not sure it was a compliment.

I was in a chemist in Notting Hill, enquiring about a pair of earplugs. The pharmacist suggested I try the ones made from foam. When I told him I'd tried them and they didn't work he said: 'My, you're a sensitive one, aren't you?'

I wasn't sure how to respond. When did 'sensitive' become 'difficult'? In an attempt to justify my request I told him I was living next door to a drummer, but I could tell he didn't care. I could have said I was a Buddhist monk travelling with the Rolling Stones and he would have had the same indifferent expression.

Growing up, I was sensitive in all the best ways. I was considerate of other children and did well in English and art. I was the friend who mediated fights between classmates. I was an only child, quiet, who read alone in my room, did origami and water-coloured. What could be more sensitive?

Then, as I got older, I became sensitive to things like criticism and rejection, weather and germs. If someone said something nasty about me, my sensitivity levels were off the charts.

And the health issues. Everything about me is sensitive in that department. My eyes are sensitive to sunlight; my allergies make me sensitive to breathing because of hay fever; I have a sensitive stomach and a sensitive palate – if I eat anything spicy, it triggers a rash on my sensitive skin. My feet are sensitive – I get blisters from walking around the house in slippers. Essentially, from my toes to my nose, I'm a ticking timebomb. I should live in a bubble. Except I'm sensitive to confined spaces.

Then, two years ago, I had a hearing test and was told that I hear decibels very few humans hear. I have dog hearing. So now I'm super-sensitive.

My superpower is to hear things nobody else is aware of. How does this help? It doesn't. Maybe if I were a spy and had to eavesdrop through concrete walls I'd be a huge success in life.

But I've always wanted to be an aficionado of something and now I am: earplugs. I know more about them than most audiologists. I've learnt that London is a great place for earplug aficionados because everyone here has a complaint about a loud neighbour.

One friend suggested I try a French brand of hypoallergenic wax plugs. She slipped me a box discreetly, like they were contraband. 'They're being discontinued,' she said, 'so use sparingly.'

Now I'm afraid to use them at all. What if they work and I can never find them again? The guy upstairs can breed wildebeest and I'll stuff my ears with cotton before breaking into my stash.

Simon swears by his earplugs. They're industrial strength and he's offered to give me a box. It seems to be a 'Welcome to London' custom I didn't know about: 'Here's an *A-to-Z* and a box of earplugs.'

The other day somebody mentioned that they had someone they wanted me to meet. Only they described him as being 'very sensitive'. No thanks. I know what that means.

# Sharing

Peole say I never learnt how to share. That's not true: I know how
to share, I just don't enjoy it.

I've always thought that I should aim to share my life with someone,
but in my experience it doesn't work out. For instance, sharing a holiday
with a boyfriend. The best part is having it to look forward to. The actual
event is never as good.

The last time that I did it, as soon as we got there I was thinking: 'I
have to remember this forever as something we've shared.' Why?
Because I knew down the line, when we broke up, I could look back on
it. Thinking about it would make me feel sad and miss him – even
though we hated each other.

Another downside to sharing an experience is that nobody ever
remembers what you want them to. There is nothing worse than saying, 'Honey, remember that time in Hawaii when I stepped on a starfish
and cut my toe?' and having their response be: 'We were in Hawaii?'

My aversion to sharing started early. As an only child, I had a bedroom and a loo all to myself. Liza is always talking about her twin sister
and I envy their closeness. But they shared a womb. Once you've shared
a womb, sharing a room is a piece of cake.

I was doomed before I was born.

When friends would sleep over, they'd touch things and move
things and bring all their stuff; it made me nervous. Nothing's changed.
When my ex moved in, it was an adjustment. He had this theory that if
you're kissing someone and don't mind their tongue in your mouth,
sharing a toothbrush is no big deal.

I didn't share that theory. Having someone's tongue in your mouth
is sexy; having their leftover broccoli isn't.

It mystified me that he had no problem moving a lounge chair the
size of a Volvo into the flat, but bringing an additional toothbrush would
take up too much space.

Worse than sharing space was sharing food. He would insist that we share the popcorn at the cinema because, as he said, it was romantic, even though what he really meant was: it costs less. At first, I agreed. Why not? I'm adventurous. But five minutes into the previews, I would reach for a handful of popcorn and the bag would be empty.

This meant either I had to speed-eat to keep up or take control and hold the bag in my lap. Which was an even bigger disaster. Because then his hand would drift over, fumble around, grab a handful, spilling most of it before re-reaching for more. And in between he would wipe his nose or sneeze in his hand – and then dig that same hand back into the bag. Nobody ever mentions that sharing food equals sharing bacteria.

I decided the only way to share the movie was to make sure we didn't talk. We could sit with our individual bags of popcorn – the same as if we were seeing the movie alone. We'd share the same air. That counts, doesn't it?

Sometimes I think how nice it would be to have somebody to share my hopes and dreams with, but then I remember: I don't really have that many hopes and dreams. So I'm not missing out.

# To Know Me is to Love Me

I met Liza's new boyfriend. Afterwards she sent me a text saying how much he enjoyed meeting me. She also said he thought I was much less annoying than he had expected me to be from my writing.

Then she wrote: 'To know you is to love you.'

Is this true? I'm not so sure.

I think to meet me for ninety minutes over sushi is to love me. After that, it goes downhill.

When people really get to know me, I've found they start to loathe me. Either that or they associate all the loneliness and isolation that has befallen their life with my influence.

Ever since becoming close friends with Sophie I've noticed that whenever there's something that's going wrong or that implies laziness or sadness – it's because she's been hanging out with me.

She used to be an intrepid, adventurous person. Now, if she has to walk four blocks out of her way to the supermarket for a diet soda they won't stock at the deli on the corner, it's an ordeal.

'I try and keep it in a two-block radius whenever possible,' she says. 'I've become like you.'

If she had really become like me, she wouldn't go out. She would get the deli man to deliver.

Another thing is that she's become a lot more pessimistic. She worries about things too. The other day she had a headache in her apartment. 'Do you think it's noxious fumes?' she asked. She hurt her knee in yoga. 'Do you think I'll need surgery?' It's not just about her either. 'I'm worried my cat might have Attention Deficit Disorder.'

I'm not sure if the angst was there all along and I brought it out or, having spent time with me, a new world has opened up. It used to be I would call her and ask how it's going and she would respond with something cheerful and upbeat. Now when I ask how things are she'll say, 'Not good.'

My next question is, 'Why?' and what follows will be an explanation of what's gone wrong or about to go wrong and then the story will invariably end with the line, 'I'm turning into you!'

I try to take that as a compliment.

People say all the time that misery loves company but only if that company doesn't make you more miserable. You have to know when to hold your ground. For instance, Sophie is very easy-going. She'll drink coffee from Dunkin' Donuts, sleep on a soft mattress, and walk barefoot on the carpet in a hotel room.

When we went away together, she was amused and referred to my issues as being 'particular'. I loved that. I'm not difficult, I'm particular.

To know me is to endure me.

# Challenges

There is a pastor in Missouri who has challenged his flock to give up whining for three weeks. Three weeks without a single complaint? That's like three weeks without oxygen.

I don't see the point. Complaining is natural. It's something that brings people together. Catholics, Jews, Hindus, Muslims, what's the common denominator? They all whinge. Even a Buddhist must give in to temptation sometimes. Chances are, there's a monk right now whispering to his friend: 'This rice is too sticky.'

But this pastor seems to think the world will be a better place if people stop complaining.

So I was curious. How long could I go for? I called Liza. 'Let's have a complaining contest,' I said. 'Whoever complains first loses.'

It wasn't a question of if, but when. 'Okay,' she said, 'but we can't begin now because I'm going to get my hair cut.' She suggested we start the following morning. But that didn't work for me, I had therapy that afternoon.

That's why this trend will never catch on. There would be nothing to talk about in therapy. Can you imagine? She'd ask: 'How does it feel not having love in your life?' And I'd say: 'It's great. I have a lot of free time.'

As the days passed, Liza and I were having difficulty co-ordinating our schedules as to when to begin. When I was done with therapy, she was going on a date. Then I couldn't start the day after that: I was meeting the accountant. Then it was the weekend: she was overbooked, I was under-booked and Sunday nights are always depressing.

There was also something else to consider: what constitutes complaining? Was it something that had to be said out loud? What if we thought a complaint? What if it wasn't a complaint as much as a statement of fact? For instance, 'I'm not happy.'

So many questions. I had to make a distinction. Complaining is not the same thing as worrying. It had to be spoken, out loud, and it had to be about a headache — literally or figuratively — or some sort of gripe.

Plus, we were each allowed one standard-issue complaint: something we say all the time so it doesn't count. Liza chose: I feel fat. I chose: I feel sick.

The day our contest was to begin, I e-mailed her in the evening to see if she'd complained yet. Her response: 'I can't remember.' She asked if I'd complained yet and I said no, but that was only because I hadn't left the house all day or spoken to anyone. In the end, the contest never happened.

It was a bad idea anyway.

# Acceptable Complaining

As levels of catastrophe have risen and become extreme, it's raised the bar for what's now considered a legitimate complaint. You can't complain about work any more without someone saying, 'At least you have work.'

Complain about your leaking ceiling and someone will say, 'At least you have a ceiling.' A few days ago I complained about my neighbours being loud. 'Okay,' my friend snapped, 'just be happy you're not in Gaza.'

Is this the barometer?

This makes me feel more isolated than ever. I still have a home and I'm not being shot at. I should feel good.

I've discovered it helps to complain about things that are universal. The senseless cruelty of nature. Global warming. And feeling fat.

Feeling fat never gets old. The 'I feel fat' complaint is ageless and not gender specific. Who doesn't love to hear someone can't button their pants? It's always welcome. Maybe because it's reassuring you're not alone while at the same time non-threatening. It's such an old-fashioned and homespun complaint. People embrace it.

If I complain about mistakes I've made no-one has sympathy. They're impatient and point out it's my own fault. But if I say I ate too much cake and feel sick? People have nothing but time. Isn't that my own fault too? Yet there's no blame. Just empathy. It's a classic.

Classic complaints have a feel-good component. 'I feel fat' is acceptable because it comes from a happy indulgence. It suggests eating too many cupcakes. It's cheerful.

Modern complaints are more unsettling. 'I feel stuck' comes from a place of emotional inertia; not getting out of bed and not caring. It suggests a lifetime of failure. This is not cheerful, it's scary.

Another acceptable complaint is discussing a bad meal. You can complain about this for hours and still have someone's undivided attention. People have no problem offering sympathy if you've had a bad piece of tuna. There's no guilt and they forget they're only on earth for a certain amount of time.

An unwelcome complaint is anything that has to do with someone's personality. Even prefacing it with 'I feel' doesn't help. People aren't interested in how you feel unless you feel fat or sick from a bad meal.

Complaints about being dumped will continue to be popular. There's always an audience for that. You can tell anyone you've been dumped and they'll feel bad for you without qualification. Although telling someone who's going through a divorce is tricky because you're likely to hear, 'Be grateful you never got married.'

I've never understood why people think gratitude is forsaken when you're complaining. There are many things in my life I'm grateful for but my gratitude is non-verbal.

# Advice

Just because I ask for advice, it doesn't mean I'm going to take it. But people get so offended. They take it so personally. It's as if there's an unspoken obligation that because they put in the time and went to the trouble, it feels like a rejection. Went to the trouble of what? Talking?

Asking for advice is one thing, being given advice, unsolicited, is another. Usually it's not what I want to hear. Which is probably why I never follow through. When I ask for advice, generally, I'm interested in having someone confirm that whatever I'm about to do or say isn't a disaster. But when I'm given advice without asking, most of the time it's completely impractical.

For instance, a few weeks ago, I was boarding a plane and over the phone I told Lori I was worried about money. 'Don't go to London,' she said. 'It's too expensive. You should stay in New York. Write a screenplay.'

When did writing a screenplay become the solution for everything?

She let out a sigh. 'Just trying to help.'

People like having solutions. Men especially. If there's a problem, it has to be solved. What's wrong with having a problem and talking about it until you get tired? Sooner or later you give up and move on.

Or you obsess about it so much you've unknowingly figured out what to do. But whenever I've tried to discuss a problem with a man he'll interrupt: 'Here's what I think . . .' before I've even finished explaining. Then if I don't take his advice, suddenly I'm being stubborn.

Relationship advice is tricky. This is when close friendships are put to the test. I never listen. If I listened every time someone told me, 'Stay away from that guy,' I'd still be a virgin.

With relationships, everyone has an opinion. Call, don't call, wait, don't wait – what does it all add up to? If it's doomed, it's doomed. It's a bad idea to give relationship advice. If a friend asks you for help and what you say works out, that will become your new full-time job. Only you don't get paid. Expect your co-dependent friend to rely on you 24/7. And if you've given advice that didn't work out, expect to be blamed for ruining her life.

No-one remembers advice anyway. I can't remember a single prescriptive thing anyone's ever told me other than a Japanese man who advised me not to eat sushi on Sunday.

My area of expertise is misery. When someone has been rejected they call me for advice. I tend to recommend: don't feel bad for feeling bad. This gives them permission to stay depressed for as long as they like. Don't rush a smile.

The worst piece of advice has to be 'Cheer up.' People tell me this all the time and there really is no response. It's useless. Especially since they always say it on days when I actually feel I'm making an effort to be positive.

# Wallowing

I keep forgetting that people have a problem with wallowing and that it's considered an undesirable trait. But why? I'd much rather be around someone who wallows than someone who lies. Frequently those are the choices. People who are comfortable with lying usually aren't big wallowers. They lack the introspection necessary for a proper wallow. Wallowers, on the other hand, are honest, but miserable.

I'd wallow all the time if I didn't want friends.

A British friend told me there's a time for wallowing. It's okay to wallow when someone dies. I pointed out that that's not wallowing; it's mourning.

I think it's acceptable because there's a time limit. People have compassion but they know that eventually you'll get over it and move on. Wallowing about life in general can be infinite. Maybe you'll never snap out and they might have to put in a lot of time sympathizing. Later on they'll mourn your death wondering why they hadn't been more sympathetic when you needed it. They'll wallow.

It's an endless cycle.

It helps to wallow about something others can relate to. Some of the more socially permissible wallows are: the cruelty of fate, the cruelty of nature, losing one's job or one's hair and animal cruelty. Nobody will fault you for wallowing about fur. However the type of person prone to wallow is unlikely to notice or care about fur. They're too busy worrying about other more pressing matters. Like how to get out of bed because life is passing them by.

The widespread wallow that everyone has patience for is the one that follows being cheated on. An extra six months of self-pity is allowed when that happens. But make sure it doesn't become a pattern; there's a difference between a one-off wallow and wallow-as-way-of-life. No-one wants to hear about it if they've heard it all before. Which is unfortunate.

If only wallowing could be seen as an admirable and attractive quality, I'd have dates every night. I'm all for wallowing and encourage people to indulge in it. Just don't talk to me about it.

# Conscientious

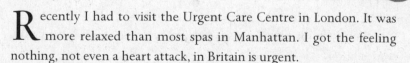

Recently I had to visit the Urgent Care Centre in London. It was more relaxed than most spas in Manhattan. I got the feeling nothing, not even a heart attack, in Britain is urgent.

I suppose the proper, dignified way to handle things would be to walk in and say, 'I'm so sorry to trouble you but when you have a second, if you wouldn't mind taking a look at this hole in my chest where I've been shot I would be most grateful.'

Two years ago I was renting a flat over the holidays and the boiler broke on Christmas Eve. I called the landlord who was in Spain and he referred me to his boyfriend who was looking after things while he was away. The boyfriend suggested I get a space heater until after Boxing Day. A few days after that the boiler man arrived but was unable to fix it because it needed a special part from Poland. A week passed. In New York, a week without heat? That's a lawsuit waiting to happen.

When I confronted the boyfriend about this he snapped, 'This isn't New York, you know. We do things differently here.'

Differently as in, freeze?

I asked a British friend — someone who knows me well — if she thinks I am pushy. 'I don't think you're pushy,' she said. 'I think of you as precise. In order for you to feel comfortable you need things to be a certain way.'

There are countless times I've been out with my British friends who refuse to 'bother' the waiter or waitress. One time I was with someone who ordered chicken and when they brought him beef, he said it looked lovely and he didn't mind, he'd eat it anyway. When I suggested he send it back he said it would make him uncomfortable. He didn't want to be rude.

With that attitude, no wonder people think all New Yorkers are gorillas. I suspect that what makes it obvious to Londoners that I'm a New Yorker is that I ask a lot of questions. For instance, 'Is that skimmed milk?' Then I'll need reassurance. 'Are you sure this is skimmed milk?' And after the waiter has taken a seat, I'll confirm. 'It's skimmed milk, right?'

I don't think that's fussy or particular or even pushy. It's conscientious.

# Neurotic

I never thought of being neurotic as a problem. In New York it means obsessing about things and worrying. Everyone in New York is neurotic; it's hardly an insult. But in Britain I've found that being neurotic is seen as a personality defect. Someone has to 'admit' they're neurotic, like owning up to a criminal record.

Until recently, I had no idea it was such a put-down. Then I got a text that was not meant for me. I had texted a friend and he wrote back saying he was out of town, he'd call when he returned. Then another text arrived. It said: 'Did I just send you a text saying I was out of town and couldn't talk? That wasn't meant for you. It was meant for Ariel, the neurotic journalist.'

When I received this I wasn't angry, I was delighted. Finally, I had the upper hand, a position I'm rarely in. So I texted back: 'You sent this to Ariel, the neurotic journalist, by mistake.' And waited. If he was embarrassed, he hid it well. He said the text was meant for his girlfriend. She wasn't thrilled about our friendship, so he'd referred to me as neurotic to allay her fears – it was deliberate. How could she possibly feel threatened by someone he called neurotic?

The explanation was good. It covered the mistake and managed to be flattering. But then it hit me: why would calling me neurotic cancel out her concern? He could have said 'unattractive' or 'obnoxious'. But apparently an unattractive, obnoxious woman isn't as unappealing.

I called my British friend Simon. 'Do you think I'm neurotic?' I asked. The fact that I was calling him with this question said it all. He paused. 'Yes. But endearingly so.' What he meant was, yes, but he has a high tolerance level for difficult people.

I wanted to understand why being 'neurotic' is a bad thing. 'What makes it negative,' he said, 'is that there's an implication that the anxiety is unjustified by the facts.'

So what are the criteria for being considered neurotic in Britain? If I'm in a pharmacy in London and I ask more than one or two questions, it's considered neurotic. In New York, it's making a purchase. It's possible the reason neurotic behaviour isn't acceptable in the UK is because Brits haven't yet come to terms with therapy. The other day I asked a British friend if she was free on Tuesday and she said: 'I can't, because Tuesdays I have . . .' Then she leant in and whispered: 'Therapy.'

She couldn't bring herself to say the word without making sure nobody could hear her. I can't understand why there is still such a stigma about being in therapy. People will disclose intimate details about their sex life but ask them if they have a therapist and they look so ashamed.

I love it when someone British does something completely reasonable and benign, then apologizes for being neurotic. For instance, calling back to make sure that I got the correct address. That's not neurotic. That's considerate.

I bet I could make a lot of money in Britain teaching classes on how to express sadness and self-hatred, or experience obsessive, paranoid and anxious thoughts. Of course, immediately following the class I'd call everyone to find out what they thought because I'd be worried.


# Having a Hook

Everyone needs a hook. Something that draws people in, holds their attention, and gives them a reason to want to see you again. Liza has a great hook: her cleavage. Whenever she goes on a date she wears a low-cut top. This usually ensures a second date. Even though she is the first to admit that her biggest insecurity is her body. She watches what she eats and works out every day so she can look as good as possible and men will be interested in her. But then, when they're interested, she worries that it's only because of her body.

But she flaunts her cleavage because it's something she can count on. No matter how stiff or nervous she might be, the cleavage will pay off. I don't think she needs it, but if it makes her feel better, why not? A sense of humour is subjective – boobs aren't.

For some people, their hook is their children. Audrey recently began seeing a man who has an adorable five-year-old son. The son is the selling point. He asks questions and seems interested in the world – much more than his dad. Plus, he likes being read to. But children grow up. I predict the relationship has a firm expiry date: his son's thirteenth birthday.

The problem is, a hook has a shelf life. If your hook is your body, you're in trouble. Liza's cleavage won't work as well when she's seventy. If the hook is power, fame or money, that fades too. You still have to talk.

I don't know what my hook is any more. It used to be my personality. When I was younger, I could be charming for hours. I could turn it on over the phone, on a date, at a party – someone once called me 'mesmerising' and they weren't even that drunk.

Now I don't have the patience to be charming for more than five minutes. I can't be bothered to sell myself and I'm unable to make small talk. Liza will spend half a day getting ready before going out with someone. If I take a shower, it's an event.

I'm lucky that my physical appearance was never part of my hook: it means I don't have to worry about the decline.

If I do still have a hook, it is most likely to be something I don't do. For instance, I don't break down crying when talking about my life. Unless I'm at therapy. I don't take heroin. And I don't whistle. Or maybe it's something I don't have. How about this: I don't have a family. Who needs another family? And I don't have any pets. No fish, no family – that has to be a hook for some people.

I'm not good in crisis situations. That's catnip for a co-dependent. Also I can be needy. My hook is for a select few.

Overcoming adversity is a hook. The more someone has struggled, the more sensitive and interesting they are. That could be my hook. Except for the 'overcoming' part. I'm still working on that.

Maybe my hook is not having a hook. If the purpose is to entice a lingering interest, I could stand out as someone who doesn't try. That's enough of a reason to see me again. It's rare to find someone so ambivalent.

# six

## SOCIALIZING

# Listening vs Talking

I love talking. Especially when I'm the one talking. There are times I don't mind listening but it's rare. If I'm listening to someone talking about me or something I'm concerned about, it's a lot easier to pay attention. Unless I'm interviewing someone. In that situation, I accept the set up: they talk, I don't. Or rather, they talk, I wait. Occasionally, I'll find it impossible to wait so I'll say: 'I know what you mean.'

If you ever have to say something I highly recommend saying this. People don't mind being interrupted with it because it suggests what they're talking about is interesting and relatable. This makes them feel good. People like to feel good when they're talking. The only downside is it encourages them to talk more.

Saying 'I know what you mean' is one thing, being asked it is another. My friend Heather ends every sentence with: 'Know what I mean?' It's replaced the pause one takes to draw breath.

For instance, she was talking about her brother and how tired she was hearing about his job and in between linking one boring point to another she said it. I wanted to say no, I don't have a brother, I'm an only child and I don't know what you mean. But that would have only made it worse. Then she'd explain.

But sometimes, you have to lie. Heather tells the longest stories. And they always involve people I don't know or care about. Someone's brother or cousin from eight years ago is staying with the sister of the friend whose mother has multiple sclerosis. She'll begin with: 'Did I ever tell you about . . .' and if I say no, she'll launch into an epic. I'll be stuck listening for the next half an hour.

Then, the last time we met she says: did I ever tell you about the time I was kidnapped? Out of habit, I was about to say yes but I stopped myself. A kidnapping isn't something you can pretend to have heard about. Some things you just have to listen to.

# Wedding Receptions

Whenever a female friend says they have something to tell me 'in person' it can only mean one thing. They've gotten engaged. In New York, everything else – including the death of a loved one and pregnancy – can be discussed over the telephone. So when Audrey called and said she had to meet, I couldn't imagine what it could be. She was married and I knew it wasn't about wanting to see me. That never happens.

We arranged to meet for coffee. As I waited for her I tried to come up with one good reason to stay. Did I really want to see photos of her new house or hear details about her promotion? But as soon as she sat down she blurted it out: 'I'm getting a divorce.' I had never even considered that. Suddenly, I had all the time in the world.

Hearing that someone else is suffering can be refreshing. It lifted my spirits. Right up until she said: 'I've never been happier.' Already, she's met someone new.

It's one thing to be happy; it's another to force others to share it with you. Whoever invented the wedding reception didn't understand this. People always say that they want their wedding reception to be just like a really fun party, but it never is. They lie to themselves and to their friends because they don't want to face the truth: eight hours of family in one room.

The best thing about a party is that you can leave any time. But at wedding receptions there's a four-hour minimum. Good luck telling the bride you're leaving before the cake. She'll look at you like you've just said the groom has herpes.

I'm not an authority on what makes a party fun, but I certainly know what doesn't make a party fun. A lead singer shouting into a microphone for everyone to 'show their love on the dance floor!' A table full of people getting drunk on champagne and toasting a lifetime of togetherness. Talking about how beautiful the floral arrangements are. Having to nod endlessly about how happy everyone looks. In a few months I have to go to a wedding in LA. One of my oldest friends, Linda, is getting married, and recently we got together in person to talk about the event.

It was the best conversation I could ever hope to have. When I'd first heard it would be in California I panicked – outside, sunshine, beachy; I was nervous about how bright it would be. But she told me the whole thing will happen in the evening in one place: a sushi restaurant. It doesn't get any darker than that.

They'll say their vows, quickly, in front of a judge and then the sushi chefs will come out with the big knives. No seating, no band, no dress code. 'If you want to wear trainers,' she said, 'I don't care.'

But then she said something that stunned me. No dancing. I stared at her. Really? She confirmed it: no dancing. It was too good to be true.

# Dinner Parties: New York vs London

I swore off dinner parties. But recently I decided to give them another shot because I was in London. And my friend Simon invited me.

In New York, 'I'm having a dinner party' means: 'I'm reserving a table for twelve at a restaurant you can't afford and we'll be splitting the cheque evenly, even though you don't drink.'

Only they never say that. I'll order a house salad and a glass of tap water and discover later that I owe as much as the person next to me who drank a couple of bottles of champagne and ate filet mignon and two desserts. Then I have to point out that I ordered a piece of lettuce not because I'm on a diet but because I'm on a budget. Suddenly I feel cheaper than an H&M handbag.

In Manhattan there is always someone who leaves before the bill arrives. They'll throw down cash, half of what they owe, and then people like me, who don't drink, end up paying even more. But if I try to pull the same trick, the hostess will shout: 'Where are you going?' And it's not like I can claim I have somewhere to go. Everyone knows I have nowhere to go.

But in London, dinner parties are in people's homes. Not only that, the guests are a diverse mix. The last time I went to one, the guests were from France, India, Denmark and Nigeria; it was like a gathering at the UN with olives.

In New York, the mix is less eclectic. It's Upper West Siders and Upper East Siders; like a gathering at the perfume department in Bloomingdale's.

For New Yorkers, talking about other parts of the world means Brooklyn and Queens. But at Simon's when I said that I'd been to Rangoon, people knew where it was. In New York, people would assume it's a trendy new club.

What's more, in London, I have the opportunity to spread the word about pessimism and encourage people to access their gloomy side. After dinner we all sat around and, one by one, guests began to reveal that they, too, had suffered moments of despair. What had been a room full of cheerful, chatty optimists became a forum for hopelessness.

Suddenly, people were confessing that they, too, had spent days in bed paralyzed by feelings of futility. They, too, had at times given up on humanity.

Then, just when I thought the evening couldn't get any better, one guest who had been holding out gave in and said: 'You're right. Life is pretty bleak.'

None of this would happen in New York — it is impossible to meet anyone there who is not already deeply unhappy. How can I depress people if they're already miserable? New Yorkers are champing at the bit to share their misfortune, and it's a competition for whose 'issues' get more air time. If you say you're in therapy once a week, someone else will say they go twice a week. Then someone else will up the ante: 'Couples or regular?' The answer is frequently: 'Both.'

It's comforting to be around that, but every so often it's good to mix it up. That's why I do so well at dinner parties in London. People love to hear from someone who complains about things other than the weather. It's so exotic.

# Parties During the Holidays

My favourite time of year is the period between Christmas and New Year. It's like a week of Sundays.

No-one expects any work to get done, the streets are empty, the pressure is off. Changing out of my pyjamas feels like an accomplishment.

Sometimes during this period I will agree to leave the house and meet up with a friend. But only if food is involved. If I'm going to interrupt doing nothing with taking a shower, there has to be an incentive. Conversation isn't enough.

That's what the phone is for. It allows me to continue to engage with the outside world without ever having to look in a mirror.

But no matter how much I try to convince people I'm happy not going out, there will always be those who don't believe me. They can't fathom how being alone and staying in is preferable to being with strangers making small talk while at the same time wondering how I'll get home.

That is always my main concern. Getting around is such a deterrent. I would go out a lot more if everything took place in my building.

A friend invited me to a Christmas party and I decided to be honest. I told her I didn't want to face the hassle of getting there and getting back. 'But you travel all the time,' she said.

Exactly. The last thing I want is to go outside my two-block radius if I don't have to.

Nevertheless, when it came to the work party, I decided to venture out. It was at a venue that couldn't have been further away. Within the first five minutes of arriving I was already putting into motion the exit plan. I got the name of a mini-cab company from the bartender and arranged for a pick-up so I didn't have to worry. But then the whole time I was checking my watch to make sure I didn't miss the ride.

Anticipating wanting to leave early, I'd booked the car too soon. For the first time in ages I was at a party I would have liked to stay at but couldn't. I had to leave because the meter was running.

On Christmas Day in London, there is no public transport. What could be better? Expectations are low. There is no obligation to go anywhere or do anything because there is no way to get around. If only it could be that way every day.

No public transport limits the people you see to whoever lives in the neighbourhood. Recently I was on the phone to someone I work with when we discovered we live five minutes away from each other. She's around – I'm around – the obligation for coffee hung in the air. 'We should meet up,' I said. And then, to my delight, she replied: 'Nothing personal, but no.'

She told me she'd rather be alone. Finally, I'd found the ideal friend to have in the neighbourhood: someone I don't have to see.

New Year's Eve is another story. If you're over the age of thirty-five, unless you're Kate Moss, no-one will question your decision to stay home. But I got the perfect invitation from Simon. The e-mail arrived with the subject heading: New Year Sadsters.

'I hereby invite you discerning types to drop in at my under-furnished flat on the 31st in the mid or late afternoon. I would arrange for there to be some food and then you could go. Or stay as you wished.'

He goes on to say he might change his mind about meeting at his flat but we could meet at a nearby pub or restaurant instead. And, we can let him know by the 30th.

What I found so pleasing about the invitation was the ambivalence. It sounded as though he was hoping we'd decline. So with the pressure off, I accepted.

Then I worried he'd cancel.

# Big Fun Parties

I never understand people who go to big parties for fun. Recently my friend Lori was invited to the kind of Manhattan party that you read about, filled with celebrities and rock stars. Everybody wants to go to those parties. Not me. Unless you're famous, those parties are never fun.

I've been to a couple of red-carpet events, and if you're not a VIP you're invisible. The last time I went to one I overheard a photographer ask a woman with a clipboard who I was. Her response? 'No-one.'

Being no-one makes being invisible sound appealing.

But just because I don't want to go anywhere, I still want to be invited. Getting the invitation is the best part of it. It's reassurance that I exist and confirms my presence in the world. Or at least that I'm still on a mailing list.

If I get invited to something, let's say an after-party for a movie premiere, here's what will happen: I'll open the invitation and feel excited. That will last three minutes. Then I'll panic. What will I wear? Will I look fat in it? I need to buy new clothes. No, I can't afford new clothes. How did that happen? I'm working all the time and yet I'm still financially in the same rut I was in a year ago. My life is a mess. What's the point of going to a party anyway? I have nothing new to say. Nothing to offer.

Then I'll change my mind. I'll force myself to go. I'll think: you can stay home any night of the week. Go, push yourself. You can leave any time. I'll talk to myself like a motivational therapist talks to a quadriplegic. 'You can do it. Yes you can. Take a shower. Don't give up.' I'll review the reasons for going. All I have to lose is a cab fare. And I might meet someone. Or have an enjoyable conversation about a new illness I've never heard of.

I'll RSVP and plan to go. Right up until an hour before. Then I'll reconsider yet again and look for signs to stay home. Is that a raindrop? It is. I'll never get a cab in a thunderstorm. Plus, what are the chances I'll meet someone anyway? I'll end up giving my number to someone I don't want to talk to and then I'll have a problem. The last guy I gave my number to at a party called the next day and left a message about how he felt our 'connection'. What connection was that? Maybe it was the moment when I asked him what time it was and then said I had to go.

Either that or nobody will ask for my number and I'll feel like a big fat loser.

And I hate networking. I don't want to be that person scanning the party for people I should talk to. I'm terrible at it anyway. If there's one person in the room who is unemployed with no connections, that's the person I'll gravitate towards.

Maybe if I drank I'd look forward to an open bar. But I don't. Sometimes watching people get drunk can be fun – but there's only so long you can stand alone before someone comes over and tries to talk.

Even though I wasn't invited to Lori's big party, I was still curious. So the following day I called her to hear all about it. 'Oh, you'll be so proud of me,' she said. 'At the last minute I stayed home.'

But I told her it didn't count – because she's married.

# Always the Plus-One

Audrey is my most social friend. As an editor of a magazine in New York, she gets invited to a lot of fancy parties and benefits.

When she was married, she didn't have to think about who to bring, because her husband was her automatic 'plus-one'. But then she separated.

This opened the floodgates for the plus-one slot. Now she rules the world. She can have any plus-one she chooses. So how come she never chooses me?

'You're never around,' she said. 'And when you are, you hate going out.'

That's true. But it doesn't mean that I don't want to be asked. Especially when it's to the benefits. A plus-one to a benefit has a lot more impact than a plus-one to a party or movie premiere. There's something incredibly satisfying about not paying hundreds of dollars and mingling with people who have.

Which is why, when I found out that she had asked Mallery to be her plus-one at the Spelling Bee benefit, I had to speak up. It was a chance to watch literary quasi-celebs humiliate themselves. Who would want to pass that up?

'I was afraid this would happen,' she said sympathetically. 'That's why I wasn't going to tell you about it.'

I'm not sure how I was supposed to be comforted by that. But then, there are so many reasons why being a plus-one isn't all that it's cracked up to be.

There's the feeling that you're an impostor – an undercurrent of anguish; knowing you're only there because of someone else's connections and clout. Whenever I'm asked to be a plus-one, it's kind of a melancholy victory. On one hand I'm flattered, but on the other, I feel like a loser. I can't help but wonder if I'm destined to always be the plus-one and never 'the one'.

But even worse than being a plus-one is being a former plus-one.

There's a hierarchy to deciding who will be chosen, and once you lose your place in the pecking order — as I have — it's hard to make a comeback.

Because the next time that Mallery has something fun to go to, she'll choose Audrey in return. Which means that Audrey invariably has others waiting — ahead of me — who she has to pay back for having invited her to something previously. The only way I'll ever get to be a plus-one now is if everyone else has the flu.

The problem is, I don't get invited to enough events and parties to maintain my plus-one standing. Which means I have to rely on people simply enjoying and wanting the pleasure of my company because I'm so much fun to be around. Need I say more?

The last time I got invited to a glamorous benefit, it was a fiasco. It was an event organized in New York, and weeks in advance I called Audrey to see if she wanted to be my plus-one. She declined. She had something else that night she couldn't get out of.

So I asked Liza. Did she want to be my plus-one? She couldn't make it either. She had to work. After that, I didn't know who to call. If you don't owe someone an invite, and you can't get a close friend to join you, it's better to go alone. Which is what I did. My only plus-one, and I couldn't fill it.

Which I guess makes me a minus-one.

# Not a Night Person

Getting older has rewards. Behaviour that was once unacceptable is becoming what's expected. For instance, when I was twenty-five and wanted to stay home on a Saturday night, everyone thought I was a loser. My friends would nag me to join them: 'C'mon, you're young! Live it up!' I tried to explain I was barely interested in living.

What makes them think I'd be interested in living it up?

But people get very offended. 'You don't like to go out, you don't like to party, you don't like to drink, you don't like to dance – what DO you like?'

I thought about this for a few seconds. 'Sitting.' I replied. I enjoy sitting. And I enjoy thinking. Sitting and thinking. Even better: sitting and thinking at home.

A decade or so later, the bar has been lowered. And yet people still need an excuse for why I'm not going out more.

Forty is a tricky age because you're old enough to get away with not going out, but not old enough to get away with not giving a reason.

I was invited to a dinner party. 'Plan to get here at 8 or so,' the hostess said, 'We can all cook together.'

I hesitated. '8 p.m.?'

She laughed. 'Yes! Not 8 a.m.!'

I began the calculations. No-one would get there on time, cooking wouldn't begin until 9, dinner wouldn't be served until 10 or 10.30. What kind of dinner party is that? We're not in Spain.

I'm not a late-night person. By 11 p.m., I'm done. I have nothing left to say, the meagre social skills I have are depleted and tolerance levels for small talk evaporate.

I can't function past 11 p.m. Even if my house were to explode, I'd still have to consider if it was really worth getting out of bed and going out.

Chances are, this isn't going to change. I can't see myself turning fifty and suddenly making a plan to meet someone at midnight. I've learned to make friends with people who hate going out as much as I do. It's hard to meet people who I have this in common with though. Where do we meet? We're always at home.

# Post-Party Regret

Whenever I do go to a social event, a few hours after it's over I start to experience Post-Party Regret. This is a syndrome that only happens to people who worry about things they can no longer do anything about.

Before going to sleep, I slowly begin to relive all the mistakes I made over the course of the evening. I lie in bed and count them the way some people count sheep. There are: Idiotic Things I Said and Idiotic Things I Did. Also: Who Hates Me.

I try to decipher which things I can let go of, which things I can let go of after a few days of obsessing, and which I can turn around to redeem myself. Then I decide who will get an e-mail, who will get a phone call, and who I can allow to continue thinking I'm an idiot.

Recently I had to create a new category. Idiotic Things I Said to the Boss. I sat straight up in bed when I remembered.

It was a large gathering and I was doing well but after a few conversations I had to sit down. I enjoy sitting alone at parties because I like to pick out those in the crowd that I am so thankful not to be stuck talking to. I feel grateful after that.

So there I was, sitting and thinking, when the boss walks by. He stops when he sees me, alone on my bench, and says hello. I stay seated. You'd think I was a hundred years old and couldn't get up without assistance.

Now I have to make a decision: is it inappropriate to discuss work? Maybe after a fourteen-hour workday, he'd prefer not to. His foot was in a cast — do I ask about his injury? It's a chance to make an impression, to say something charming or funny or perceptive.

'So,' I said, 'have you ever heard of Wegener's disease?'

I'm not sure this was the best way to go. He looked confused. 'You've never heard of it?' I say. I remember going into way too much detail about this rare autoimmune disease and, as his eyes began to glaze over, a voice in my head said: 'Stop talking. Now.'

But I was in so deep I had to save it. Quickly I tried to make a connection — were there any presidents, prime ministers or movie stars who had Wegener's? But it's so rare, it hasn't even got a celebrity who has it.

I wish I could go back in time.

Maybe if I'd enquired about his injury I might have discovered I'd had the same one. Or tennis — why didn't I talk about that? I love Billie Jean King. Why didn't I bring that up? We could have become buddies. Bonded for life. We could be talking about tennis right now. But no, I had to spend five minutes with the boss talking about a disease. And people wonder why I don't go out.

# seven

## MODERN MANNERS

# Cut to the Chase

Cut to the chase! When did it become okay to interrupt someone and say this? Why not just say, 'I don't feel like listening any more.' or 'Get to the point because I'm bored.'

Everything is being sped up. Everyone wants the bottom line and brevity. Wasting a second of someone's time is criminal.

What's everyone doing that's so important?

If I send someone an e-mail longer than five sentences I'm worried it will take too long to read and they'll give up. The three words no-one wants to hear are: 'Let me explain.'

When someone tells me to cut to the chase, I usually do. I'm better off. And also, naturally, I assume I must be rambling and going into too much detail. Or maybe the person I'm talking to has somewhere they need to be and they don't want to miss the ending of what I'm leading up to. I try to be positive.

But there are women out there who men would never tell to 'cut to the chase'. Can you imagine Brad Pitt and Angelina Jolie sitting down to breakfast, she starts talking about Darfur and he tells her to 'wrap it up'? Angelina could talk for eighteen hours without drawing breath and no-one would mind.

I've been in plenty of situations where someone is telling a story that is dragging on but I would never order them to cut to the chase. It seems rude. Why do that when you can simply stop paying attention?

Especially if you're talking on the phone. How hard is it to interject 'Really?' and 'No!' while doing dishes or making the bed. That's the considerate way to treat people.

The other day I began telling a friend about the results I got back from the doctor and she told me to cut to the chase. 'Fine,' I snapped. 'I'm going to live.'

Is this what it's come to? Have attention spans become so truncated that any verbal communication has to be reduced to bullet points?

I admit that sometimes I can get carried away. My father once said, 'When I ask for the time, I don't need the history of Switzerland.' But now, I've lost all perspective. I was out to dinner with someone and he asked what happened with a recent break-up. I asked him on a scale of one to ten how much detail he wanted. 'Four.'

Four? That seemed impossible. But fine. I proceeded. And then, just as I got going he began to wave his finger in the air in circular motions giving me the 'wrap it up' gesture.

I was giving him detail level four and it was still taking too long? 'Forget it,' I said.

To make matters worse, I've always been drawn to the cut-to-the-chase type of man. When people ask me what type I'm attracted to I should say: the type who will rush me to get to the end of a story so that he can go back to what he's doing. Like taking a nap.

I went out with someone once who would set an egg timer. I'd have to get to the point before the time ran out. I felt like an athlete in training. The kind of athlete who didn't care about winning.

People love to point out it's not the destination that matters but the journey. So how come that never applies when I'm telling a story?

# We Are What We Do

Where's the dividing line between what we do and who we are? It's never more of an issue than when I'm in New York. People there are obsessed with work. Your job is your identity. The first question anyone asks is, 'What do you do?' And it's not just what you do for a living that matters but where you do it. If you call someone at an office the secretary will ask for your name and as soon as you give it the next question will be: 'From?'

I freeze when this happens. Will my response get me through? My friend Linda always replies, 'Texas.' For those of us who don't work for a company, there aren't a lot of choices. Where am I from? The Ariel Leve Foundation. We're based in my apartment. It's a small operation.

The other day I had to call someone in publishing. 'From?' the assistant asked.

'Nowhere.'

That did wonders for my self-esteem.

In Los Angeles, everyone wants to be in show business so it's a given the day job is temporary. You know your waitress is waiting to be an actress, the postman is writing a screenplay and the dry cleaner is also a producer. It's understood.

But in New York, it's far more aggressive and people are scrupulously judged by what they do. If you say you're a writer, you'd better be prepared for the next question: 'Who do you write for?' Or if you're an artist: 'Which gallery?' Or a musician: 'Who are you signed with?'

If you don't have an answer then you're made to feel as though you have no right to claim this as your profession. People at parties in New York are like the IRS. Slip up and you'll get audited.

And don't bother starting a sentence with: 'I used to work for . . .' because unless the story ends with: 'And now I own . . .' whoever you're talking to has stopped listening and moved on.

Here's the difference between New York and London. If I tell someone in London I have a book coming out, they say: 'That's great. Congratulations.' In New York the response is: 'When's your pub date?'

My what? At first I thought it was a mistake. I don't date at pubs. Then it was explained: the publication date – when the book hits the shelves. This was explained to me by a housewife who I met in the laundry room. As in the actual laundry room, not the name of a hot new club.

Another thing I've noticed is that writers in Manhattan who have been published can sometimes be stingy about allowing unpublished writers to claim 'writer' status. As though there are a limited number of slots. My favorite distinction is, 'She is a published author.' Without the word published in front of it, it means: wishful thinker.

I've decided that in the future I should introduce myself as 'A published author who feels like an unpublished one.'

Maybe the dividing line comes down to what we feel is most familiar and how we see ourselves. In which case, I could refer to myself as a chronic worrier.

## Don't Ask Where She Got It

Women in New York are secretive about all the wrong things. A total stranger will be more than happy to share all the intimate details of her overactive bladder with you, but ask where she got her boots from and you've gone too far. She'd sooner give you her ATM pin code.

I went up to a woman in a restaurant who was wearing a blazer I coveted. I excused myself for interrupting and politely inquired where she bought it. I explained it was exactly what I'd been looking for, for years. She stared at me. Her eyes narrowed. 'Europe,' she said, defiantly.

Not one to back down, I pressed on. 'Really? Where in Europe?' She fired back: 'Paris.' I raised a quizzical eyebrow. Then she gave herself away by adding: 'I don't remember where I got it.' And that's when I knew my instincts were right: she was lying through her teeth.

What does she care if I own the same blazer? She'll never see me again.

But it's Manhattan. It has trained women to be suspicious. Her thinking was: who knows what could happen? First I could steal her blazer, then I could steal her job, then her boyfriend – it's a slippery slope.

If you approach someone, whether it's an acquaintance at a party or a stranger on the street, and ask where they got something, you're likely to hear a range of replies. 'London' is a very popular answer. Also: 'It was a gift.' Or: 'I found it on a bench. In a park. In London.'

I have a friend who, when asked where she bought something, always says it came from the Salvation Army. 'Even if it's Marc Jacobs,' she says. 'They never know the difference.'

There are women who will tell you they stole a jumper from a lost-and-found box in a church rather than reveal the store where they purchased it. Why? I've thought about where this comes from. For a lot of women, getting dressed is a form of expression, and when somebody copies their look it's like fashion plagiarism.

There are lawsuits for stealing words – why not for style? Luckily for me, that's one thing I'll never have to worry about. Nobody would want to steal my style.

In New York, if anyone ever tells you the name of the shop or the actual designer, chances are they're from out of town. But in London it's different. Women are willing to tell you where they bought something because it's always the same answer: Topshop.

It doesn't matter if it's a bag or a pair of shoes, a blazer or a belt, and yet whenever I've gone there to follow up, the merchandise has changed dramatically.

This leads me to believe Topshop is the dental-floss lie. As in, everyone tells the dentist they've flossed, but nobody ever does.

There are so few resources left that identify us as unique, people feel a need to keep the most mundane things to themselves. Recipes, for instance. All great cooks have their secrets. But a lot of mediocre cooks do too. I tried some guacamole at a party once and told the hostess how good it was. She shot back: 'Thank you! It's got my secret ingredient!' What, avocado?

I wanted to tell her I didn't care, I don't cook, and it wasn't even that good — I just needed something to say. But I didn't. And when she commented on my dress and asked where I got it, I told her I made it.

# Compliments

Someone once introduced me as 'world class'. I don't know what that means. Who decided 'first class' wasn't good enough? But unless 'strange-odour detector' follows 'world class', the description doesn't apply to me.

My response was: 'I'm not.' Suddenly, everyone's expression changed. I could tell they were revising their decision to talk to me. That's because in New York the inability to take a compliment is a sign of weakness. People believe that if you put yourself down, there must be a good reason. And they take it personally if you don't accept their compliment.

Audrey says her failure to take a compliment is something she's working on in therapy. So whereas in Britain dismissing a compliment is modesty, or charming self-deprecation, in America it's a syndrome.

Compliments are tricky. The first time someone in London told me I was 'brilliant', I was flattered. Until I heard her use the same word to describe a roast chicken.

Then there are the back-handed compliments. For instance, 'She's an intelligent woman. For an American.'

There are also the compliments that leave you wondering how to respond. Women experience this when someone says: 'You look thin.' I'm never sure whether to say thank you. Thinner than what? Than the last time you saw me, when I looked fat?

A friend of mine who is heavy hates it when she's trying something on in a store and the saleswoman tells her: 'That looks flattering on you.' She knows what she's really saying is: 'That's our last size fourteen: grab it.'

I used to be sensitive to criticism about my work, but working in the UK I've had to adapt to a low-level praise threshold. In the beginning I'd fish, but that didn't come across well. The more I asked, 'Do you think it's good?' the more 'constructive' criticism I got. Now if someone says, 'Well done' I'm thrilled. So maybe that's the trick. The fewer compliments you give, the more they matter.

That doesn't necessarily apply in relationships. I dated someone who was very affirming and generous but then I discovered it wasn't just with me. I told him it didn't mean anything when he called me sweetheart if everyone else was sweetheart as well. Complimenting a woman is never easy, but here's one for men to avoid: 'You have a great vocabulary.'

Telling me I'm the smartest woman you've ever met doesn't work either. If I'm the smartest, the others must be pretty dumb.

Luckily, it doesn't happen often. In the same category would be when someone says to a single woman: 'You're such a catch, how is it that you're not taken?'

It never makes us feel better about ourselves. All it does is remind us the world is filled with people who think we're special, who we're not with. From now on, if someone asks how it is that I'm not married, I'm going to reply: 'I don't know. Maybe I'm unlovable.'

The most frequent compliment I get has to be: 'You're not as neurotic as I thought you would be.' Now that's a compliment I can accept.

# Second Opinions

I'm all for second opinions. And not just for medical matters. Second opinions in general are a good idea. Especially when they confirm what I want to hear. Then when they don't, I disregard them. Where's the down side?

In order for a second opinion to be truly effective you have to believe someone knows more than you do. I was with a friend in a store and she tried on a coat. 'I like it,' she said. 'What do you think?' I told her I thought it wasn't great. Hearing that, she put it back. This made me uncomfortable. I shouldn't have that much influence. What do I know? I'm not a fashion stylist.

Unless you're an expert, I'm not sure how much a second opinion should count. A second opinion from a neurosurgeon on whether to get brain surgery? Take it on board. A second opinion from a friend who's cranky and ready to go home? Not so much.

Also the subject you're asking for an opinion on determines how candid someone will be. 'In my opinion you shouldn't have crab cakes' doesn't carry as much responsibility as: 'You should get a divorce.'

There's a difference between a second opinion and an unsolicited opinion. A man from the shade company came to my flat the other day to measure. I was explaining to him that I wanted a cord with a cleat to tie it. He made a face and suggested I get the chain. I told him I liked the cord.

'I prefer the chain,' he said. I stared at him. 'That's nice,' I said. 'But since we're not getting married, I'll take the cord.'

In spite of the fact that everyone's got an opinion there are certain things I've found it's difficult to get a second opinion on. For instance, when I bite into something that tastes funny. I'll hold out my fork and say, 'This is disgusting. Does this taste rotten to you?' Not many people will offer to help with that.

I'm always looking for second opinions. Did you hear that?

Do you smell that? Does it smell weird? Generally the answer is a resounding: no.

Not that it matters. I'm not the kind of person who is easily swayed. This could be because I assume that when someone is giving a second opinion, they're not really putting a lot of thought into it. Especially when it comes to relationships.

Here's a question no woman will answer honestly: 'What did you think of him?'

Friends say what they think you want to hear. But I can tell when a friend doesn't approve of someone because they'll only mention the way he looks.

'He's cute.'

And?

'Tall.'

And?

'Good posture.'

Only when it's over will I find out what they really thought. 'I knew right away he was a feckless, selfish, delusional, Communist.' I love it when that happens. But then sometimes, I get defensive. Just in case we end up getting back together.

If, like me, you're an indecisive person, a second opinion will invariably lead to a third opinion. Which leads to a fourth opinion to split the difference. And if the fourth opinion creates a tie, a fifth opinion is needed to break it. It's a lot of work.

The things in life I'd really love a second opinion on, I'll never be able to get. For example, when a man says: 'I missed you.' Does he mean it? If only I could ask his therapist for a second opinion.

# Apartment Stalking

I spoke to my friend Sophie who had just returned from an afternoon of apartment stalking. It's a little like deer stalking. Only instead of taking place in the woods, it takes place in Manhattan.

In Britain there are plenty of apartments for sale but there are no mortgages to buy them with. In Manhattan the problem isn't finding the money, it's finding the right apartment. Or rather finding an affordable apartment that comes with sunlight. And New Yorkers have highly developed strategies and skills to get what they want.

My friend Jane got her two-bedroom by befriending the doorman of the building she wanted to live in. Every time she passed by she would stop and chat — until one day he said the woman in 14-C had died. Before the obituary ran, she'd signed the lease.

I was in college when I first heard of apartment stalking. A guy I liked would make daily trips to the building he wanted to live in and hang around outside. Just being close to it made him happy. One day he asked if I wanted to join him. I stood outside on Central Park West in a snowstorm because the way he felt about the building, I felt about him. Eventually it sunk in: he would never be as excited about me as he was about the apartment.

New Yorkers also know how to navigate the newspaper real estate section. It's similar to reading the personals; you have to decipher the code. Great lobby means the apartment's a shoe-box. Northern exposure means from 4–5 p.m. you'll get sunlight. I met a broker who asked, 'Do you really need light?'

When I asked why she explained it had really good closet space. You can't have everything.

Sophie has been living with a friend on the Upper West Side for the past year while she searches for her perfect place. She's become obsessed with finding The One. Every week or so a real estate broker will let her know there's something new on the market and she will tell me she is on her way to see it. Just like a blind date. She describes the location, the neighbourhood, and the size. Expectations are high. I can hear the excitement in her voice.

Half an hour later she will call with the bad news. She sounds depressed and hopeless. The more this happened the more she began to explain them as bad dates. One 'didn't speak English.' Last month it was a 250-square-foot studio in Chelsea for $2000. She walked in and walked out. 'He was 5'2" with a comb-over and smelled of stale beer.'

Then she found true love on 15th Street. She had been to visit a friend who lives in the building and was told another similar apartment was opening up on 1 September. She called me that afternoon. 'I know where I'm going to live,' she said. 'It was meant to be.'

Suddenly she started telling people her new address. 132 West 15th Street. And she would frequently say it out loud – testing it out – the way women will dreamily recite the last name of the man they want to marry. She would write the address down and draw a heart around it.

She began sentences with 'My new apartment' and made trips to the area to scope out her future dry cleaners, pharmacy, and supermarket. When we happened to be walking on 15th Street and Sixth Avenue she pointed to a nail salon. 'This is where I'll get my mani-pedi.'

Now for the past few weeks she has been stalking the Super. Every building has one, a superintendent, who manages the property on site. Like every good restaurant has a maitre d'. She calls, texts and e-mails him constantly. This Monday, he's agreed to meet. Meeting the Super is like meeting the parents. She got a manicure.

Just then it occurred to me to ask if she's ever actually seen the apartment.

'Not *if*, but *when* I get the apartment, I'll make it work,' she announced.

There was no room for doubt. Plus, standards had been lowered. As long as it had running water and a front door, she'd be happy.

She had decided this is where she was living. Just like an arranged marriage. I only hope when she meets the Super he doesn't tell her that he's already committed her true love to someone else. With deer stalking the objective is to end up with a dead animal. Apartment stalking, the objective is to end up with a place to live — not a dead Super.

# The E-Me

I am an entirely different person over e-mail, and the e-me is far more appealing. With e-mail, I don't have to listen. In a phone call I repeat myself, so I'm easily found out. The person I'm talking to will say: 'You just asked me that! Weren't you listening?' Then I'll have to be honest and say: no.

Also, the e-me is brief. A few typed sentences and I'm done. But if that e-mail had been a phone call? Two hours. No wonder my friends all e-mail me. If I call and leave a message for them to call back, they'll respond by e-mail.

It used to be plans were made over the phone and broken over the phone. I preferred it. When I leave someone a message about a plan we've made and I see an e-mail from them an hour later, I know what's coming: a broken plan. On the phone, they might hear disappointment in my voice. Nobody wants to deal with that. But with e-mail I'll write back: 'Don't worry about it.' No guilt, no interrogation, no accountability. It seems so unnatural.

People enjoy it when I'm brief but the feeling isn't mutual. Especially when it comes to relationships. I ran into an ex-boyfriend and we had an awkward moment. So that night, rather than call, I sat down and wrote him an e-mail about our relationship, what went wrong and why. I laboured over every word, every comma, making sure it was just right, revising it until it got to the point where it was so perfect it was publishable. You know what he wrote back? 'Okay.'

When it comes to one-word responses, not answering at all is a lot better. At least then there's some mystery. But writing 'Okay'? Why not write: 'It's sad that you put so much time into writing this e-mail when, clearly, I don't care.'

I suppose the problem with misrepresenting oneself over e-mail is that, eventually, people will think you've changed. Then, when the true personality comes out, it's a let-down. I had a fight with someone over e-mail, and when she realized I wasn't giving up and, sadly, I was still me, she called. 'I got tired of typing,' she said. 'Let's just do this over the phone.'

With friends, it's one thing. But with people you don't know, it can be tricky. Liza flirts freely in texts whereas in person she's more self-conscious. Last week she got a text saying 'Yummy'. She had no idea who it was from, but she wrote back 'Busy' because even though she didn't know who she was sending it to, she didn't want to shut it down completely and risk offending someone who might just be a future husband.

I'll never send sexy texts because I like people to lower their expectations of me, not raise them. I don't do sexy e-mails either. Reading someone's e-mailed sex talk is like reading poorly written erotica. If there's going to be any sex talk at all, it has to be over the phone. That way, when I'm not listening, nobody will know.

# Gifts

Honesty is never a good idea when it comes to gifts. No-one wants to hear it. The last time I was honest about a gift I received, it was a mess. My boyfriend had spent a lot of time picking it out and said he knew it was something I'd love. Right then I knew I was in trouble. The difference between a gift that's a hit and a gift that bombs comes down to expectations. As soon as he told me I'd love it, I prepared to be disappointed.

Everyone knows it's the thought that counts. Unless it's from someone you're sleeping with. What they choose is really about how well they know you, where the relationship is going, and when it will end.

A fluffy magenta cardigan with burgundy velvet heart buttons doesn't say: I'm thinking of you. It says: not a good sign. Was he drunk when he saw it? Maybe it was intended for someone else. An ex-girlfriend. Or his gran. What could ever have led him to believe I would wear magenta? If he paid attention to me when I wasn't naked, he'd have noticed that I wear black. No magenta, no fluff, no velvet heart buttons.

'What do you think?' he asked as I held it up. But before I could answer, he added: 'It was really expensive because it's one of a kind.' There were just so many things wrong with that statement, I didn't know where to start. One of a kind means un-returnable. And why tell me it's expensive? I'm not Victoria Beckham.

Gift-giving often reminds me that people in my life have no idea who I am. Last year, my friend Audrey gave me a halo that lit up. Which would have been great, if I was nine. And the other day I found a beautiful notebook my father had given me a decade ago for my birthday. Inside he had written: 'The life of Ariel Leve 1998–2008'. I flipped through it. Empty.

I've always had a problem giving gifts as well. The last time I went to a baby shower, I gave the mother-to-be Primo Levi's book *If Not Now, When?* I forgot that at baby showers people sit around and watch as the gifts are opened. When a book about Jewish partisans fighting Nazis was unwrapped, everyone's face fell. But quickly an enormous box from Pottery Barn was shoved in front of her and the mood picked up again.

I'm terrible at giving gifts because I always get people what I'd want. If I could, I'd get everyone I cared about a free MRI scan. I'm also a slow giver. It once took me so long to buy a friend a wedding gift that by the time I sent it off, she was getting divorced.

When I told the boyfriend I didn't like the cardigan he seemed so much more hurt than he had a right to be. I'd hurt his feelings. Why should that hurt his feelings? It's not like I was saying he smelled bad. I was saying that something he chose didn't work. You'd think he'd appreciate knowing this. He didn't. His response? 'I'm never giving you a gift ever again.' And he didn't.

# Phones on a Plane

The only thing worse than being seated next to a crying infant for a transatlantic flight is being seated next to a crying infant whose mother is talking into her mobile before takeoff exploring all the possible explanations for the tears.

'She started four hours ago. No, her nappy is dry. Yes, I burped her. Wind? You think it might be wind? Okay, I'll tell her it's you.'

[Pause] 'Sweetie, Daddy's on the phone. Here. On the phone. Talk to him. It's Da-da. DA-DA. You don't want to talk to him? You don't want to talk to Dada? [Pause] She doesn't want to talk to you. I'll call you when we land.'

Recently, I was horrified to discover that airlines will soon allow midair mobile phone use. I read a report that said, 'At last, certain flights on Emirates Airways are allowing passengers to officially get their calls and texts.'

Are people that busy that they can't go a few hours without talking on the phone? And, if they are really that important, shouldn't they be able to afford a private jet?

The first flight that authorized calls was between Dubai and Casablanca. There was one main topic of conversation: how weird it was to be talking mid-flight. 'We've taken off and I'm calling you from the plane! Isn't that weird?'

But soon, the novelty will wear off and there will be a series of other topics. For instance:

Topic 1: Reception. 'I only have two bars. I can't hear you. I SAID: I CAN'T HEAR YOU. CAN YOU HEAR ME? I SAID: CAN YOU HEAR ME?'

Topic 2: Location. 'We're somewhere over the Pacific. I don't know where exactly. I'll call you back when the clouds break.'

Topic 3: Confidentiality. 'Yes, no. Yes. Uh-huh. Can't talk. [Whispering] There's someone sitting right next to me.'

Topic 4: Torpitude. 'Where are you? You're in Boots? What shampoo are you buying? Don't get the dandruff shampoo, it made my scalp itchy.'

Topic 5: Grievance. 'They took away my jar of Crème de la Mer. It was 4.2oz. Oh, and they took away my nail clippers. And something else. Oh yeah, my tweezers. Make sure there's a pair of tweezers at the hotel when I land.'

Topic 6: Drama. 'I didn't catch last night's episode of *Desperate Housewives*. What did I miss? Go on – I have loads of time – start from the beginning.'

Topic 7: Tragedy. 'You think it's malignant? Has it metastasized?'

Topic 8: Therapy. 'I can make it a double session. I'm having panic attacks again and I need to go over my abandonment issues.'

Topic 9: Help. 'Are you ready? First take the printer out of the box. I'll walk you through it. Now plug it in.'

Topic 10: Rejection. 'You're dumping me? Now? While I'm stuck on a plane?'

Then there's the potential for phone sex. I'll have to wear a biohazard protection suit on all flights.

Whatever happened to having to turn off all equipment because it's dangerous? An eight-hour transatlantic telephone conversation from one airborne bore to another is a recipe for air rage. And will there be a distinction between calls made in economy and those in business? Because I'd sooner downgrade than sit next to someone strategizing about marketing manoeuvres.

Also, the list of my questions at check-in is going to get a lot longer. I'll want to know if the person seated next to me has a whiny voice. I'll want to know if they are chatty and if their mobile is fully charged. And what about the pilot? Will he be chatting on his mobile too? That's a little unnerving.

When they find the black-box recorder after the crash we'll discover his last words were, 'Hang on, I'm losing you.'

# eight

## OTHER PEOPLE'S RELATIONSHIPS

# Marriage

I've been thinking a lot lately about marriage. It started when I ran into an old friend on the street.

'What's been going on?' She asked.

We hadn't seen each other in a long time. 'Oh, you know . . .' I said, trailing off. This is a great thing to say. The unspoken part of that sentence could mean I'm on top of the world and being modest – or maybe it means I've had a life filled with wrong turns and bad choices which I don't feel like sharing. I'll let her decide.

She stared at me. Or rather, she stared at my hand which, to her dismay, was concealed in a glove. Then she cut to the chase.

'Are you married?'

I could have won the Nobel Peace Prize and it wouldn't have mattered. 'Nope,' I said, 'not married.'

She made a sad face. 'We have to get you married!'

We do?

I know she intended it in a caring way, but it implied I must be unhappy because I don't have a husband.

There are a number of reasons why I'm unhappy but not having a husband isn't one of them. Which, now that I think about it, seems uncharacteristic. How did I miss that? Surely if I had a husband it would open up new opportunities for disaster.

Most of my friends whose weddings I've been to are no longer together. Yet you never hear anyone say, 'We have to get you divorced!'

It's not that I'm against marriage, I just haven't cared one way or the other. And with friends who are happily married, it was definitely a desired preference.

For my friend, Lori, the symbolism as well as the language was important. They had been living together for a while but the words 'boyfriend' and 'girlfriend' seemed juvenile.

'Now when he calls me his wife,' she says, 'it's shorthand for the person I chose to spend the rest of my life with and the only person I'll ever kiss with tongue again.'

I can see the merits in that. People say all the time that when you get married you give up your freedom but that's not necessarily such a bad thing.

I asked my formerly married friend, Audrey, what freedom she'd had to give up and she replied: 'Thinking just about myself. When you're married, you're always considering the other person.'

That sounds like a lot of work. She said it was. For instance, when she was single she would get asked to things and she would only have to think about what she felt like and her schedule. When she was married she had to factor in the time she had with her husband and if it was something they could do together. Was it a party she could bring him to? Or a movie he'd want to see? And so on.

That wouldn't be an issue for me though. I don't get asked out to a lot of things so I wouldn't have to worry.

What people tend to forget is that being on one's own is often by choice. The other day Sophie told me someone said she was 'brave' for not being married. She took it as a compliment. I can see how leaving war-torn Sarajevo with nothing but the clothes on your back is brave. But living in a one-bedroom in the West Village without a husband – is that really heroic?

'It is,' Sophie said. 'Because a lot of people get married not to be alone.'

I guess I always assumed that even if I got married, I'd still feel alone. But maybe I wouldn't. Maybe I'd feel so attached to my husband that instead of panicking about something happening to me, I'd panic about something happening to him. That would be worse. I can see it now: finally I find my soul mate and a month later he chokes on a fishbone.

If I got married it wouldn't end in divorce court but at the Dignitas clinic for assisted suicide.

Then again, that's kind of romantic. At least we'd be together.

Aside from having someone to help you die, I think the best thing about marriage has got to be telling people. A few years ago I got a call from an ex. We hadn't spoken since the split and he called to say he was getting married.

'I wanted to tell you myself,' he said, 'before you heard about it from someone else.'

That's nice. But how would I have heard about it from someone else? We had no friends in common. Did he think it would be announced on the news?

Perhaps because I was the last person he went out with before deciding to commit himself for life to another woman he felt he owed it to me to make it clear that when he said 'something was missing', it wasn't with him.

Then he said, 'I only hope that someday you can be as happy as I am.'

What do you say to that?

Recently I ran into him at a party. By now, all emotion had neutralized and I was interested to meet his wife. Turns out, they're no longer together and he seemed pretty miserable. So I guess not only am I as happy as him, I might even be happier.

Who'd have thought that was possible?

# Wandering Eye Guy

Liza is currently using a Jewish dating agency. Even though she's ticked 'no' to keeping kosher, going to synagogue and observing Shabbat, she still feels it's better to date within the tribe. Her reason? A lot of Jewish men live on the Upper West Side, and she likes to stay close to home.

A month ago, she found someone she liked, but in the photo he e-mailed, he was jumping out of a plane. That would have been enough for me to say: no way. But Liza enjoys the adventurous type.

The following night they met in a bar and hit it off. She noticed something was wrong with his eyes. She couldn't tell what because it was dark, and besides, she liked him, so it didn't matter.

The next date was dinner. When she arrived, he was already waiting. The restaurant he chose was, like the bar, dark. Over a candlelit meal she tried to figure out what it was that didn't seem right. Afterwards, they stood on the street and, yet again, she couldn't tell. Just as she was about to propose that they get on the brightly lit subway so that she could get a good look at his face, he hailed a cab. I told her the only thing left was to go on a daytime date. For Liza, this is tantamount to a date without clothes.

It takes her about eight hours to achieve date-worthy hair, and going out in the afternoon meant she would have to wake up at dawn to get ready. If you call her during the hair process you're likely to hear: 'Can't talk, doing hair.' Putting the product in, sitting completely still for an hour and then sticking her head in the freezer to lock in the curl. The end result is the 'natural' look.

I admire her devotion. If my hair is clean, I consider it done. If I have a date then I'll pull out all the stops and brush it.

She took my advice. When he suggested the cinema, she chose a matinee performance. Finally, she would get to see his eyes while queuing for tickets in daylight. But when he showed up, he was wearing sunglasses. And he kept them on right up until the movie began. By this point, she was really beginning to like him. Whatever was wrong with his eyes wasn't important any more.

A few nights later, he called. He would be unavailable for a while because he was getting eye surgery; he'd call when it was done.

'Eye surgery?' she asked. 'For what?' He told her he had a wandering eye that he was having corrected.

A few weeks went by. And a few more. She left a message on his machine saying she hoped things had gone well with the surgery and she was looking forward to getting together again. She never heard back from him. Ever.

'I guess I was only good enough for him when he had a wandering eye,' she said. 'Now he probably thinks, "Hey, I'm a Jewish man in New York with two good eyes. I can get Natalie Portman."'

# Future Husband

The wedding vows in the *New York Times* are required reading for women in Manhattan. 'Barbara loves wearing masks to parties and met Joe when she dropped her programme at the opera and he picked it up.' Liza loves reading the vows. Not me. I've always thought there should be 'Divorces' on the opposite page. 'Barbara and Joe thought they'd found true love but then she caught him cheating on her with the interior decorator.' Who wouldn't want to read that?

One evening I called Liza. She said she couldn't talk, she was getting ready for a wedding. She gets invited out a lot. She's pretty, friendly and great at small talk. People like Liza.

After she and I hung up, her phone rang. A male voice said 'Hello,' and even though she was unsure who it belonged to, she didn't ask 'Who is this?' in case it was someone she knew. She doesn't like to offend anyone or hurt anyone's feelings. How did we end up friends?

The man asked what she was doing. 'Getting dressed,' she said, still not sure who she was talking to.

'So what are you wearing?' he asked.

She assumed it was her friend Jim and replied: 'A strapless dress.'

'What are you wearing underneath?' he ventured. Casually she responded, 'A bra and underpants,' and expressed concern about visible panty lines. He interrupted: 'Are you wearing La Perla?'

This made her pause, but only momentarily. Jim is gay: it's possible he would ask this. 'No.' She doesn't own any La Perla.

'Would you like to?' Now she knew it wasn't Jim. Jim is cheap. He would never buy her La Perla.

'Who is this?' she asked. The man replied, 'Steven,' and sounded embarrassed.

They laughed, and Liza was hopeful. Maybe this was her future husband. She could see the vows announcement: 'They met by accident when he called the wrong number. She was on her way to a wedding.' She continued chatting.

'That was a hot conversation we were having,' he said. He enquired if her bra was a push-up. A strapless, she answered, and the questions quickly became more provocative. He asked if she liked kinky sex, if she liked being kissed hard or soft. She thought about it and responded, 'In between.' To stay neutral.

Then Steven asked her the craziest place she'd ever had sex. When she told me the story, I couldn't believe she was still on the phone with him. But she wanted to come up with something exciting. It might have been a dirty phone call from a total stranger, but she didn't want him to lose interest.

Finally she said she had to go. 'Just leave me with one hot erotic thought for the night,' he says. Now she's stressed. She tells him she likes being waxed, and says goodbye. Immediately she knew she could have come up with something better.

So an obscene phone call leaves her feeling bad that she wasn't dirty enough. Then I remember something even more disturbing: why is it that when I called, she didn't have time to talk?

# Surprises

Growing up, my friends loved surprises. Not me. Surprises in my life were never a reason to celebrate. Surprise! The nanny quit while you were at school. Or surprise! Daddy lost custody – you're heading back to New York.

Surprises were disappointments waiting to happen. When it comes to gifts, I've never understood the thrill of opening something without knowing what it is first. You're likely to be underwhelmed and how do you hide that?

The only good thing about a surprise gift is that it's over in a matter of seconds. With a surprise party, the torture goes on for hours.

Whoever came up with the idea of a surprise party was obviously wanting to punish someone. Who doesn't want to know ahead of time that they're about to walk into a room filled with friends who are all decked out. There you are in your sweatpants while everyone else looks their best. How is that fun?

No good can come of a surprise. For instance, I had a boyfriend who used to meet me at the airport. I got used to it and looked forward to it. Of course, it wasn't technically a surprise because I knew he'd be there, but I was still always astonished when I saw him because I couldn't believe he cared enough to have made the effort.

Then it stopped. I got off the plane one day and he wasn't there. He told me he tried, but it wasn't possible. Why wasn't I surprised?

The riskiest surprise has to be a blind date. Liza went on one recently. Her first thought on seeing him? Okay-looking, tall but a little dorky, with screwed-up teeth. Or, as she put it, 'He didn't make me sick.'

At what point did the standards drop so low that not being made sick is a good thing?

He told her the date itself was a surprise. He said to meet on the corner of 79th and Fifth Avenue at 6 p.m. Then he informed her that they were going for cocktails at the Metropolitan Museum of Art, and dinner reservations were at 10 p.m. Ten? She called me from the toilet.

'I'm not sure I want to spend one hour with him, much less four,' she said.

They had drinks, got a cab and headed downtown to a restaurant she never would have chosen in a million years. Too fancy, too many courses. Then back uptown and – surprise! More drinks. By the time they got back to her house it was 2.30 a.m. He could have taken her to Italy.

When I asked her how she finally got out of it, she said she had realized that if she didn't kiss him, it would never have ended. It turned out he was a great kisser. Really great. Surprisingly great. He still has screwed up teeth, he annoys her, and he doesn't make her laugh – but now she's worried he won't call.

Maybe he'll surprise her.

# Foot Rubs

My friend Heather has the ideal situation now. She is going out with a man who is besotted with her feet and wants nothing more than to give her a foot rub every time they meet. They haven't had sex yet, and the last time I met her for dinner, he kept sending her texts about what he wanted to do to her feet. It wasn't pervy. It was about massaging her arches. What could be better?

After dinner, Heather said she was going over to his house – for a foot rub. Like a footy call, only she was worried it would inevitably turn into a booty call, which would be a shame. All that time wasted on plain old sex when she could be having her toes caressed. I told her not to worry. Chances are he wasn't interested in sleeping with her: her feet might be all he desired.

But that just worried her even more. 'What's wrong with the rest of me?' was her immediate response. And there was something else to consider. What if she was leading him on? Was she a toe-tease? Flashing her feet? Spicing things up with sexy nail varnish in spiky heels? The situation was becoming complicated.

It had started innocently enough. One night they were walking and talking and he invited her over for a glass of wine. They sat on his sofa and he said she should put her feet up. So she did. She happened to be wearing stockings, so she wasn't particularly concerned about the condition her feet were in. It began with him gently touching her feet and led to a full-on foot job. 'It was very erotic and unexpected,' she told me, 'but it didn't lead anywhere and I left.'

A few days later, she was on her way back from the chiropodist. She'd just had a cortisone injection in her heel. She was in his neighbourhood, stopped by and, when she got upstairs, slipped off her shoes and pointed to the plaster. It was the first time he'd seen her bare feet and he was enthralled. 'Oh, your feet are so cute!' he said. An ex-boyfriend had once said her feet were 'exquisite'. So 'cute' was a let down.

After that, she played it cool. She put her shoes on and they went to dinner, but back at his house he was much more appreciative. 'Give me your gorgeous feet,' he said. Not only that, he said she has 'young-looking' feet. That did it. She surrendered.

She says she knows it will end, but for now she can't give him up. He's better than the people at Happy Feet – a place in Manhattan that specializes in foot rubs. I'm jealous. I walk more than she does. And my arches are higher. I go to the chiropodist too, and not once have my feet ever got a sympathetic cuddle afterwards. Now they've become callused and cracked, nobody pays attention to them — unless it's during a pedicure. And that doesn't count, because it's like having a foot-hooker. They're not really into you, you don't know who they were with an hour earlier and you have to pay them.

# Infatuation

Audrey is in the throes of infatuation with a new man. They have strong e-mail chemistry, and over the weekend they exchanged six e-mails, four pages long, single-spaced. It started in the context of work, then one day he asked her something personal. She asked him something personal in return, and the flirting had begun. Very quickly it became sexual, and now, even if she wanted to stop, she can't: the momentum is too exciting to resist.

'The last time I was single was before e-mail,' she told me. 'Where does this type of relationship go?'

Here's where: down.

I've been there, I know. I told her it's a simple equation. Life is like jail. The only way to get through it is to have a diversion and e-mail is the courtyard where you can breathe some fresh air. First it's the occasional flirty one back and forth, then it's checking obsessively to see if there's a new one waiting, then you're meeting in person, and the next thing you know you're in therapy six times a week, eating peanut butter in your bathrobe and wondering why your friends won't talk to you any more.

Infatuation isn't real. That's why it feels good. If only we could see through the haze of intoxication and lust, we'd save time and cut right to the misery. Then we wouldn't have to deal with the memory of how great it was in the first six months when everything was bliss. It's the bliss part that's so hard to forget.

Recently a US census declared that the average marriage lasts seven years. I couldn't believe it. That long? My longest relationship was six years and it felt like an ice age had passed. The top three reasons given for divorce were:

Reason 1: Partners were not good matches. I blame this entirely on infatuation. When you're infatuated, you want to go to bed with someone all the time, which is blinding, and you stay blind until the third year, when you open your eyes, see who you're really in bed with and realize you'd rather go to dinner. That's when your partner says something along the lines of, 'Again? We go out all the time,' and you decide: eyesight sucks.

Reason 2: People change. I blame this too on infatuation. Because it's not that the partners change, but your focus. You see them for who they really are and have that 'what was I thinking?' moment. That moment is followed with an addiction to *Oprah*.

Reason 3: Partners either hid their psychological problems or they developed over time. No, it's not that they hid them, it's that when you're infatuated with someone, the things that would be unacceptable in the real world become tokens of endearment and affection in infatuation world. 'He's not a psychopath, he's just attentive.'

As for psychological problems developing over time? That's nature telling you infatuation has worn off. Knowing all this, I tried to deter Audrey from getting sucked into the e-mail vortex, but I didn't sound very convincing. Why? Because infatuation is a break from reality – an extreme, head-spinning, heart-thumping heartbreak waiting to happen. It's the only time you don't see what's coming and there's no voice in your head pointing out the future. I'm jealous.

# Operating Instructions

Audrey's e-mail relationship has progressed to actual dating. He has been sending her texts letting her know how much he likes her and this is making her anxious.

She is not used to someone being so direct – in fact she finds it a turn-off. The other day she said she wished she could tell him not to text her until after 7 p.m. Then she could spend the entire day worrying that he'd changed his mind and gone away. If he did that, it would be a huge aphrodisiac.

I don't have that problem. Whenever someone says they like me I don't believe them and don't trust it. But only if I like them too.

Wouldn't it be great if men came with operating instructions to maximize their performance and shelf life?

For instance, with my ex, here's what the instructions would have said: Warning: emotional conditions may render appliance subject to frequent disconnections and inoperable. Best used in short bursts over limited periods of time. Machine will regularly go into pause mode. This is not a malfunction, just standard operating procedure.

And the warranty would have a long list of what's not covered including reliability, accountability, responsibility or magnanimity. And any cost for repairs will not be refunded.

Sophie's operating instructions would read like this: Congratulations on your new purchase. Unfortunately, this machine's original instructions were misplaced during manufacture so you'll just have to figure it out as you go along. Due to a systems failure at age six, unexplainable glitches may occur at random moments, especially when machine is dealing with actual or imagined threats of abandonment.

My own current operating instructions would be more complex. To begin with, the pamphlet would be so extensive, no-one could cope with reading it.

But like an iPod — it would include a step-by-step illustrated Quick Start.

However once you began following it, it wouldn't make sense. Step 1 wouldn't take you to Step 2. It would skip to Step 4 and say you had to guess what Step 3 was.

And there would also be a note that said: Damaged in transit — irreparable — it is not purchaser's fault.

The trouble with handing someone an instruction manual is that most likely, they'll ignore it. So maybe a list is more manageable. I asked Audrey to make a list of what a guy should do to make it work with her.

1. Be emotionally available. But not too emotionally available.

2. Don't talk about ex-girlfriends except when I'm hounding you for information.

3. If you must talk about them only respond with disparaging remarks.

4. Don't invite me to dinner Tuesday night on Tuesday morning. But don't ask me out for the third Friday of June 2011 either.

5. Make me laugh.

6. Be sensitive. But not so sensitive that I worry that I might make you cry.

7. If I get up to go to the bathroom after sex, don't say it's too bad I have to go home so soon.

8. If you're going take a break from the relationship, tell me face to face rather than disappearing with no explanation.

9. Ask me about me once in a while. Even if it's just once or
   twice a month.

10. Make me laugh.

Here's my list:
   Show up.

# Faking a Share

I was on the phone with my friend Emily talking about a private matter when suddenly I heard a voice in the background ask who she was talking to. 'Ariel,' she said.

I didn't know that her boyfriend was in the room. 'I thought you were alone,' I said. You would think there could be some advance warning.

Does talking to her mean that I'm talking to him? It does. And even if she tells me he's not listening, that's not the point. Most people don't listen. But I want to choose who they are.

What I really worry about, though, is the pillow talk. That's when I assume she tells him everything. All the secrets I prefaced with 'This is just between you and me' are shared because when you're in the pillow-talk zone, you lose all sense of discretion.

In relationships, pillow talk is prime time. Crucial information is exchanged and the dangers are immeasurable. Lying face to face, you throw caution to the wind and open up. It's like confession, only without the Hail Marys. What's the worst that can happen?

The next day, you remember the worst that can happen: he can tell someone. I tend to regret things I've shared during pillow talk, but once it's out there it can't be reclaimed. I walk around in a haze of sharing-shame. Why did I tell him that? Who will he tell? The only thing I can do is hope that he wasn't paying attention. Which is likely.

I always over-share. How I'm feeling, what I'm thinking . . . people don't need to know. Audrey under-shares. She was telling me that she was with her new boyfriend and they had just had sex. He started opening up about his past relationships. She listened, knowing soon it would be her turn. There was an unspoken obligation that she too would have to reveal something. She couldn't just say, 'Oh, that's nice,' and roll over. So what did she do? She fake-shared.

I've done that before. Men don't catch on. I'll reveal something that isn't exclusive to pillow talk but they think it is and feel closer. For instance: I have abandonment issues. It suggests I've disclosed an intimate secret when really, I'd tell this to a taxi driver. I rationalize it's okay because chances are he won't remember it anyway.

The accidental share is the one that I really dread. This is when I go one detail too far when talking about a past relationship. I can tell by the look on his face that he's reconsidering everything while simultaneously plotting his escape.

I've decided that, from now on, I'm not going to share. Which should be easy, since I have nobody to talk to.

# Bartender Crush

Liza has a crush on a bartender called Ed. 'Watching him open a beer was a turn-on,' she said.

I didn't see the appeal. So she put it in terms I could relate to: 'Imagine watching someone sew up a wound.' I got it. When I asked what was the exact moment she fell for him, she said it was when he bought her a drink. 'Here,' he said, refilling her glass, 'the next one's on me.' Liza is easy to please.

It's sexy to watch someone work when they're good at what they do.

She found out the nights when he's bartending, and now she's there all the time. That's fine, because it's a bar. But if it were me, it would be a little more challenging. My options are limited. I don't drink, so a bar is out. I don't go to clubs or work in an office. Where would I go? A&E? Not the best place to hang out.

I suppose I'm most likely to meet someone in a doctor's waiting room. But it's hard to develop a crush on someone when I'm preoccupied with the disease I'm about to be diagnosed with. The one time I struck up a conversation with an attractive man sitting next to me, it didn't go well.

I pointed to my arm and my opening line was: 'Do you think this rash is contagious?' He switched seats.

The best thing about Liza's relationship is, she's in complete control. She only goes into the bar when she thinks she looks pretty, and never has to worry about making a plan. There's no pressure or hassle of having to cancel, because he doesn't know when she's coming. The downside is there's a TV above the bar that is constantly tuned to a cable sports station. Now she has to feign interest in the baseball game. Whenever her (new) team loses, she has to remember to look concerned. It's hard work. She also has to remember which team she's rooting for.

The other thing is, because he's a bartender, he keeps filling her glass. It gets to the point where, after the fourth refill, when he's not looking she'll pour some of the wine out into the water glass. Plus, he could retire early on the tips she's leaving him. But she's having fun. So who am I to point out that she's about to become a broke alcoholic?

Of course, the best thing about a crush is the possibility of something happening. Having it become a reality could be a disaster. If Liza asked him out, two things could transpire. He could say yes, and she would lose her control over the situation. Or he could say no, and she'd have nowhere to go when her hair looks good. Either way, it wouldn't be a crush any more.

She wanted me to meet him. I thought, why not? It would be fun, just like the old days. Except now we're old.

When we got there I told him I didn't drink, so I'd have water, and that I'd be eating but I was off wheat. And before I could explain, he said: 'You're off wheat? We have a gluten-free menu.'

I thought he was great. But Liza was upset. 'He knows more about you in two minutes than he knows about me, and I've been coming here for three weeks, drinking myself into a stupor.'

I told her she needs to get some issues. She's not neurotic enough. The good news is, finally, there's something that I can help her with.

# nine

## LIFE IN THE SHADE

# Philosophy

Everybody needs a philosophy. A simple line you tell yourself that instantly curbs the desire to say or do something that you'll regret.

I dated someone once whose philosophy was: never say never. Or, as I saw it: create false hope. Nothing was ever definite. And anything's possible. He'd swear he didn't want to get married again. Then he'd add: 'Although, who knows? I never say never.' What he really meant was: 'I may get married again, just not to you.'

Emily had a great philosophy: enjoy the moment. The problem was, she kept pushing it on me. I have enough trouble living in the moment. I have to enjoy it too? But she was determined, and the more she expected to enjoy the moment, the less it happened. So now she has a new philosophy: no regrets.

That's the beauty of having a philosophy. When it doesn't work, you just adopt a new one.

I don't know what my personal philosophy is any more. It used to be: question everything. But that was exhausting. Plus, it annoyed people. Besides, it's already my nature to be suspicious. I don't need a philosophy that makes life more complicated.

Next I tested out live and let live. If someone did something that bothered me, I reminded myself to let it go. Unless they broke a plan at the last minute. Or kept me on hold while they took another call. Or forgot to return an e-mail. So much for live and let live.

I've gone through a bunch of philosophies. Most of them didn't stick. I'm currently trying out: do unto others as you would have them do unto you. The other day was a good test. My friend Simon called in a panic. He left a message on my voicemail at 7 a.m. 'You will get an e-mail that says "For Amanda" in the subject line,' he said. 'Please do not open it. It's private.'

I sat at my computer and stared at the unopened e-mail. It would be so easy to take a peek – he'd never know. What did it say? Who was Amanda? What could have provoked this mild-mannered British man to sound so terrified? The thought of being emotionally vulnerable had him gripped with fear.

I began to think of reasons he wouldn't want me to see it. Maybe he confessed he was a spy. Or gay. Or even better: a gay spy. Why wouldn't he want me to know that? I wouldn't care. It had to be something else. Something I would care about. Maybe it mentioned me. It was excruciating and I was desperate to read it. But I called him back and promised I wouldn't open it.

'That doesn't mean someone else couldn't read it,' Liza said. Meaning I could have forwarded it to her, let her read it, tell me what it said, and still have respected his wishes. I considered this option. But then I thought how, if it were me who had inadvertently sent an e-mail, I would hope that a friend wouldn't open it. So I clicked delete.

Of course, if it did happen to me and I was told that it had been deleted unread, I'd wonder if they were telling me the truth. And therein lies the one philosophy I always return to: trust no-one.


# Cheer Up

There is nothing worse than telling someone to cheer up. It is futile. Not only that, it's damaging. There is an unspoken obligation. Cheer up. Or else. For the person who has received this mandate, not only do they feel bad, but now they also feel bad about feeling bad. It's a lot of pressure.

All my life people have told me to cheer up. I'll be walking down the street or waiting for the light to change and suddenly someone, a total stranger, will call out: 'Cheer up!' Often followed with: 'Hey, it's not that bad!'

How would they know? Maybe it is. Maybe it is THAT bad.

I was in an elevator recently when a woman said, 'Why the long face?'

I replied: 'Why not?'

She paused before insisting it's not as bad as I think it is. I wanted to tell her I found out my entire family was killed in a car crash. That would show her.

But I didn't. I let her think I was scowling unnecessarily.

Over the years, I've had a variety of responses to when someone tells me to cheer up. Sometimes, if I'm feeling generous, I'll fake a smile. That always makes people feel better. 'That's more like it,' they say and I hold the fake smile until they've disappeared.

Other times, I'll simply ignore them and continue to scowl. They look disappointed with their inability to get through to me, but keep going.

If you don't want to hear about a problem that's fine, say so. But 'Cheer up' offers nothing. Saying nothing would be preferable. At least the silence could be interpreted as sympathy.

I've developed a new technique for responding. I discovered it when I had just gotten into a taxicab and was looking forward to staring out the window in silence, thinking my unhopeful thoughts. The cab driver looked me over in his rear-view mirror and said: 'Hey girlie, cheer up.'

I looked right at him and stated: 'No.'

For a second, he didn't know what to do. I worried he might pull over and tell me to get out. Would he argue and insist I cheer up? As it turns out, he took no for an answer. He simply shrugged and said, 'Okay.'

His total indifference cheered me up. Nobody really cares anyway.

# Escape Fantasy

There are times I'll be walking down the street and think: 'If only I lived in a house in the middle of nowhere — I bet then I'd be happy.' I'll walk along thinking this, comforted by the idea that, at any moment, I can leave my city life behind for a rural, peaceful one.

A while ago I visited a friend who lives in the countryside of northern England. Before I arrived he told me the house had no doors. 'You mean it's a loft?' I asked. No, he said, he ran out of money and the builders left early. Nothing has happened since, except that he's lived there. This made me nervous. A house on the moors with no doors? It sounded draughty. But there are times when one has to take a leap of faith. I decided to go anyway. I'd have an adventure.

But adventure isn't always a good thing — I need to remember that.

My friend's house is made of millstone grit, a stone quarried from the Yorkshire moors and used for pavements in London. I don't mind pavement when I'm standing on it, waiting for the lights to change. But as a kitchen floor, surrounded by walls made of giant slabs, it was prehistoric. I felt like I was visiting Fred Flintstone.

I was introduced to a man who waved hello while holding a bucket of eggs he'd collected from the chickens living inside his house. That can't be sanitary. Suddenly my neighbour in New York with the smelly cat didn't seem so bad.

My friend told me that his house was built in 1836 and, as we climbed the stairs, I could tell that some of the original dust had remained untouched. He pointed out that the house had never been cleaner.

In honour of my visit, he'd gotten rid of 'most' of the cobwebs, and washed and hoovered the walls next to the bed. Hoovered the walls? What sort of creatures were living in the gaps in the stone that would need to be hoovered out? Just then I noticed a framed giant black beetle.

As we reached the first floor, I gazed at the ceiling: there was a hole in it the size of a foot. I was told the floor had woodworm and in certain spots it was weak. 'It shouldn't be a problem,' my friend said. 'You're small and light.' I felt better already.

What woman doesn't love to hear she's thin enough not to worry about falling through the floor?

So the wood had worms, the walls had nests, the only thing missing was mice. 'Oh, I haven't seen one since I put the poison out,' he said. Turns out that what I thought were chopped fresh herbs near the breadboard had been green pellets of Alphakill. Good to know.

I would be sleeping in a room where the window had just been opened for the first time in over a year. It was like camping indoors. I piled on my clothes, stuffed cotton in my ears to stop anything from crawling in, and went to the loo-with-no-door to brush my teeth. That's when I was handed a screwdriver. Hot water was a treat.

My friend doesn't want new taps: he wants the old ones restored. But the parts from salvaged taps are in short supply, so he's lived for two years turning the hot water on from under the sink. Either that or skipping it entirely. You never think about how much you appreciate a tap until you don't have one. Same with a door. My trip to the country was brief but meaningful. Only now, I've lost my escape fantasy.

# Behind My Time

Most people think they're ahead of their time. Not me. I was born three decades too late. If only I had been born during the Great Depression, I would have thrived. Everyone was miserable, unemployed and worried. Does it get any better?

Here's why I would have loved living in the 1930s: buying an apple was seen as an achievement. I like that. Today there's way too much pressure on having potential. Back then — no-one would care if I wasn't fulfilling my potential. Someone could ask, 'What did you do today?' and I could respond, 'I baked a potato'. That would be enough.

Recently there have been surveys that show that in the twenty-first century we're materially richer but emotionally no better off. Maybe if I were materially richer I wouldn't mind. I'd drive my Mercedes out to my beach house, listening to my iPod and wondering why surveys exist in the first place.

But given that I have all of the stress of modern-day life and none of the perks, I find it only confirms I was born too late. At least if I was materially richer, I'd have an excuse for my misery.

Research also shows that living in the 1930s, life was simpler. I can see why. I'd be much better off owning one dress, one pair of shoes, one lip-gloss, one phone and one pair of sunglasses.

Plus today, life expectancy for women is an excruciating eighty years. But back then, life expectancy was a more manageable sixty-three. Who needs to live longer than that? Once you hit thirty, you're middle aged. I'd love it.

From everything I know about living during that era, there was a lot of anxiety about day-to-day life. And people pulled together. Fascism . . . poverty . . . disease . . . I could complain all I want and no-one would find it unusual. I'd have good reason. Now if I complain about a disease — people just tell me to go to a doctor. Back then, I could sit on the stoop all day and no-one would care. Expectations were low.

# Old on the Inside, Young on the Outside

People say you live and learn, but sometimes that's not the case. Sometimes you just live. And keep going. Or what you do learn you forget.

There are so many unreliable people in my life that when someone does something normal, it's a big deal. Like when someone says, 'I'm just being honest with you,' and I feel grateful.

How did it get to the point where honesty deserves admiration?

Every emotional lesson that I've ever learnt in my life, I've forgotten. There's no growth, no inner peace. Just anticipation of what's to come. The rejection I feel now when someone dumps me? Exactly the same as when I was fifteen. Worse, even. Because now I factor in that I should know better.

There are times when I'm waiting for the light to change and I'll see someone with a scar across their face and think: I shouldn't feel bad. It could be worse. I could have a giant scar across my face. For about three blocks, I'll walk along appreciating what I have. But later that night, looking in the mirror, I realize it's still me. And the woman with the giant scar isn't going to be standing next to me at a party next time I'm feeling invisible, to remind me what's important and boost my self-esteem.

A lot of elderly people I've met tell me that, even though they're seventy on the outside, they still feel like they're twenty on the inside. I find this disheartening. Feeling twenty on the inside doesn't sound that appealing. Does this mean that at seventy, I'll still be sitting by the phone wondering why someone hasn't called me back? Forty years from now, am I going to hear myself saying: 'What bingo game? No-one told me about a bingo game,' and feel left out?

There is no comfort knowing that as I get older, I'll still feel youthful. I'm living and learning but to what end? This means that all I have to look forward to is me the way I am now, but with wrinkles and thinner hair.

# Wristbands

I'm drinking a lot of water lately. But no matter how much I drink, it doesn't help. I called Dr Samuelson and said: 'I can't get enough water. I'm worried. It could be diabetes.'

'You don't have diabetes,' he said. 'You're thirsty.'

Shouldn't he test me?

'But I'm drinking all the time and I'm peeing all the time.'

He said I'd be fine. But how does he know?

I didn't have the chance to mention the spot on my lower back that's itched for the first time ever and has turned dark brown. It may be a melanoma.

'Are you sure it isn't a freckle?' Liza asked. We were out to dinner. Sitting next to us was a man wearing an orange rubber wristband.

I wondered what it was for. 'Whatever it is,' Liza said, 'I deserve it more. How about a support bracelet for being forty, single and alone?'

There's a reason why there isn't a Single and Alone wristband. Nobody would wear it. I leant over and interrupted the guy's dinner. 'Excuse me, what's the orange wristband for?' Leukaemia. Okay, we felt bad after that.

There are rubber wristbands in every colour, all over the world. It can get confusing. In America, a blue band is in support of ovarian cancer. But in the UK it has the words 'beat bullying' on it. I thought it was a statement to stop bullying people with ovarian cancer. Then I discovered it was a separate cause.

Lately, it seems a lot of people are wearing what goes best with their outfit. I saw a woman with half a dozen wristbands on her arm. How does she know where to concentrate her support? How does she know what to wear with them? And if you wear the wristband every day, it says you're behind the cause, but the moment you take it off, then what? People will think you're no longer supporting leukaemia.

There's another dilemma. What if you become attached to a particular bracelet and want to wear it all the time? Imagine the disappointment on reading that a cure had been found.

Now there are wristbands for everything, from favourite football teams to political candidates. But where are the wristbands for depression?

Since wearing a wristband creates a groundswell of support, I'd like to wear one that goes direct to the source. Me. My wristband would be for raising awareness of a dire situation: mine. It would be black and it would say: Hopeless. It would sanction wallowing. But it would have to stand for passive support. I wouldn't want anyone going on *Oprah*. I don't want advice. I just want to wear the band, quietly, in solidarity. It goes with everything.

Turns out there already is a black wristband. It's for melanoma.

# Things a Pessimist Should Try

When people make a list of things they'd like to do before they die, do they ever consider they might die tomorrow? I consider this all the time. Not that it motivates me.

As far as I'm concerned, if you're going to follow a list, it might as well be things you have a personal stake in. I tried to come up with a list of things I cared about doing before I die. I didn't get very far.

I'd like to have one week where something doesn't go horribly wrong. But that's more of a wish than an activity. I persevered and came up with a list tailored to my personality.

The problem is, any list I come up with will be met with the question: what's the point? Still, there are things every pessimist should try at least once. Even if it's only to confirm you were right to think it wouldn't make a difference.

Things a Pessimist Should Try before They Die.

1. Get out of bed with an excitement to start the day.

2. Consider substituting 'Great' for 'I've been better' when asked: 'How are you?'

3. Go for a walk on the beach without worrying about skin cancer.

4. Fly without demanding a seat next to the emergency exit.

5. Wait for your luggage without panicking it's on a flight to Guam.

6. Ride in an elevator without listening for the grinding sound that tells you it's stuck between floors.

7. Touch the pole on the Tube without pulling your sleeve down to cover your hand.

8. Sleep in a hotel bed without inspecting the mattress first for bed bugs.

9. Buy a lottery ticket.

10. Don't assume that the waiter gave you regular instead of decaf.

11. Use the nights you have insomnia to knit a scarf rather than ponder every mistake you've ever made.

12. Get through at least one week of 'four weeks to a better body'.

13. Attend a wedding without thinking they'll be divorced in five years.

14. Give someone a week to respond to an e-mail without deciding they hate you.

15. Go to a party without expecting to hate it.

16. Tell a friend: 'You'll get over him' even if you don't think she will.

17. Show up at work without worrying you're about to lose your job.

18. Go out in the pouring rain believing a cab will turn up in seconds.

19. If he tells you he's busy and can't talk don't assume he's having an affair.

20. Have someone explain why the stock market always corrects itself.

21. When someone tells you: 'This too will pass' don't ask: 'But what if it doesn't?'

22. Look at an old photograph of yourself without thinking your best days are behind you.

23. Just once, experience what it feels like to say: 'Life is good.'

# ten

## SWEATING THE SMALL STUFF

# Recycling

I received a card from Royal Mail informing me that since I wasn't home (which I was) I would have to go to the post office to pick up a package. How exciting. A surprise gift.

I woke up the following morning to get there early. Anticipation was high. What could it be? Energy saving light bulbs from British Gas.

Do they send these out to every single customer who has electricity? Or just those they suspect have a disastrous carbon footprint? Also it must cost them money. I'd rather they lower my bill and I'll buy my own light bulbs.

A few days later, a friend was visiting. He entered my kitchen and saw the light bulbs still sitting in the box. Then he saw a roll of paper towels.

'Don't you care about the planet?'

Of course I care. I recycle. I turn off the tap when I brush my teeth. And I walk everywhere. I don't own a car or a house and I've never flown in a private jet. Can Prince Charles say that?

Last year there were over 300,000 private jet flights in and out of London. I wasn't on one of them. I don't know what my carbon footprint is but I bet it's smaller than Al Gore's. And yet, if I ask for a plastic bag at Waitrose, I'm made to feel I'm personally responsible for killing the polar bears.

Why is it the most disapproving looks at the checkout come from mothers who have baby buggies and four-wheel drive Porsche Cayennes idling outside with the nanny at the wheel?

I feel bad that maybe I'm not doing enough. But then it occurred to me I'm doing more than most. I'm not having children. That's about as environmentally friendly as it gets. Putting fewer people on earth does far more to prevent global warming than buying organic blueberries.

Not washing nappies at home or using a nappy washing service means I'm conserving water and electricity. Also, children love baths. They get very dirty. It takes gallons of water to fill a tub. No children, no drought.

There's no car-pooling to school, no eight-hour return journeys to the country house every weekend, no bicycles or tricycles, which means less factory production emissions, less packaging, distribution costs and less landfill as they grow out of them every three months. No driving from one shopping mall to another to find the Barbie dream house, which is definitely not biodegradable.

And holiday flights. I won't have to cut back on flights to Disneyland since I'll never have to go. Children drink milk and can grow up to eat meat. Cows, pigs, and sheep are a huge source of methane gas. Plus transporting animals uses water and fuel — think of all the energy it takes to get that protein to the supermarket. Even worse, they could grow up to be meat-eating millionaires. Taking a Gulfstream jet to Paris for steak frites.

Less consumption of food, less bottled water, less tumble drying, fewer polystyrene cups, and plastic knives, forks and spoons. Fewer iPods, digital cameras and laptops. Not to mention the tons of toilet paper I'm saving.

Not only am I protecting the environment, but I'm protecting innocent people from coming into a world that is being destroyed. Why show them the Grand Canyon only to follow with, 'And one day it will all be gone.' How cruel is that? I'm not just an environmentalist, but a humanitarian too.

Of course, this is all subject to change. I could meet someone tomorrow, have children, and my carbon footprint will become the size of a sasquatch. But chances are, that's not going to happen. I'm more likely to adopt.

Adopting a child is a form of recycling, right?

# Technology

What's happening to the world? I used to be content to sit at my desk and work on my computer. Now everyone I know has a camera built in so they can video-conference. I feel I'm missing out on something that I don't want to be a part of to begin with.

Whenever a new gadget comes out I get this sinking feeling, because it's yet another area of life where I'm going to be left behind. Technology is constantly moving forward; I'm not.

Men in particular have gone gadget-crazy. The more a man is obsessed with a gadget, the less attractive he becomes. If I'm interested in someone and he pulls out a BlackBerry, he might as well be pulling out a copy of *Dianetics*. Part of it is that I feel resentful. That BlackBerry is getting all the attention that should rightly be given to me.

But also, when a man I like has a BlackBerry, this means that access to him is unlimited. He's able to phone, text and check his e-mails anywhere, and there's never a reason why he can't be reached. So with all that at his fingertips, how is it that I still haven't heard from him?

I liked it the way it used to be – when I had room to assume there were valid reasons he hadn't called back. Any number of things could have happened. He could be out, in transit, or just not at his desk.

But now, when someone has the ability to check their messages while walking down the street? There's no excuse. If he can be available at all times, then the only reason he's not available is because he doesn't want to be.

Another reason I find being around people with BlackBerries disheartening is that it's a reminder that someone else has so much more going on than me. They have so many friends and so many plans, they have to manage them continuously.

BlackBerries contain people's schedules, and they're always checking to see what event they have to go to. There's nothing worse than sitting next to someone while they tap away and look up every five seconds to say: 'I'm almost done . . .' My response is always: 'Don't worry, take your time.' It's not because I'm being considerate. It's that I don't have anywhere else I need to be and no-one is waiting for me to get back to them.

Whenever a friend gets a BlackBerry I go through a mourning period. It's like they've gotten married or become born again. I know even though we're still close, it will never be the same. It's like they've moved to Brooklyn.

# Giving In

When I first met Sophie she was holding her BlackBerry so tight it was clear it was a situation. Right from the start, I knew what I was in for. She didn't try to hide her attachment or pretend it was only used for work. She was upfront – friendship with her meant I would be receiving 'sent via BlackBerry' e-mails.

I adapted and got used to it. Soon I discovered I even enjoyed it. I liked that I could always reach her and that I would receive instant responses to all of my questions. There was no end to the conversation.

Then one day she told me she felt there was an inequity. I could reach her at all times but she couldn't reach me. When, she asked, would I be getting a BlackBerry? I put it off. And made excuses. But I knew that if I didn't give in our friendship would suffer.

I got an iPhone instead. Only I was so freaked out that it stayed in the box for a month. Then slowly, I began to test it out and see how it worked. I would practise for a few minutes and use a new feature. Each day I could tolerate using it a little bit longer. Like an athlete in training, I was training for my sport: making phone calls.

In London, my mobile-addicted friend Simon was impressed. His phone dependence had gotten so bad that his family are considering an intervention. So using the iPhone around him felt a bit cruel. Like ordering cake with someone on a diet.

But Sophie was thrilled. 'You have no idea how happy this makes me,' she wrote back to the first e-mail I sent from the iPhone.

I was less enthusiastic. Partly because I was confused. I would send her messages that asked: Is this working? What time does it say it arrived? UK time or US time? What time did you get it? Why is there a delay? Figuring it out was a full-time job. And there was a sense of defeat. The point of it was to make my life easier. Maybe my life isn't capable of being easier.

People say all the time that technology has improved our lives but I'm not convinced. Using my laptop computer has given me backaches, eye problems, wrist-aches and a sore neck. There's also the potential for getting RSI from using the keyboard, and getting a hump. No-one should have a hump before they're ninety.

I got a call from Sophie. She was depressed. Her BlackBerry wasn't working and it would be five days before it was fixed. She repeated this with disbelief: 'Five days?!'

I was nervous because she seemed unable to cope. All emotion in her voice had been neutralized. She wasn't sleeping. 'I'm waking up every 15 minutes checking to see if it's working again,' she said. 'I feel like Tom Hanks in *Castaway*.'

That she felt cut off from the world on the Upper West Side of Manhattan was disturbing. But even more disturbing was that I understood. It's not that I need to send e-mail at 4 a.m. – just knowing I can is a comfort.

In the end, it wasn't five days. It was two days. And it was over a weekend – not a lot of e-mail traffic. When she let me know it was working again she sounded ecstatic. The way most women sound when they announce they're getting married. 'I'm so happy I could cry,' she said. I'm not so sure this is a happy ending.

# It's Good Enough

For two whole years, scientists, supermodels and designers were working for Victoria's Secret to come up with the Ipex bra. When it was launched it was called the 'world's most advanced bra'. Is this bra really a scientific breakthrough?

Shouldn't scientists be focusing on more important things? Lung cancer is still a problem. Then again, I suppose if you're a scientist and you have an opportunity to collaborate with Gisele, that's hard to pass up. The lab at Victoria's Secret has got to be more fun than the lab at Johns Hopkins University.

What were the think-tank sessions like? Scientific minds pursuing the phenomenon of expanding cleavage and the gravity of boobs? It's not like they can experiment on rats. Also, what took two years to figure out? Given the amount of research, this bra should be able to act as an early-warning system. Two years? Einstein's theory of relativity took less time to develop.

In the ads, there were photos of Gisele with nothing on but the bra and studious-looking spectacles. This must be because everyone knows you have to be an intellectual to wear a wireless bra. Someone was sitting in a marketing meeting one day and exclaimed, 'I know! She'll wear glasses! That will show how scientific it is.'

I don't know anyone who shops at Victoria's Secret any more, anyway. They should take some of the money they're putting towards 'research' and put it towards making a bra that actually lasts more than one laundry cycle. Not that men care about any of this. I've done my own research and have found the only thing men really look for in a bra is that it comes off fast and easy.

Being a scientist is rough. Your discovery has to be useful and catchy.

What do we know about Archimedes? Pi. Galileo? The telescope. Marie Curie? Radioactivity. All the good discoveries have been used up. If scientists really want to come up with something that helps women, here's an idea . . .

A few weeks ago, I went for my very first mammogram and the experience couldn't have been more unpleasant. The room was air-conditioned, so it was like disrobing inside a meat locker. The technician's hands were no better.

'Did you just dip them in ice water?' I asked. She wasn't amused. The machine itself was like a torture device from the Middle Ages. If they can invent a way to heat a towel rail, you'd think someone could find a way to heat a mammogram machine. And while they're at it, a heated speculum.

I'm not saying advances aren't important, but does it ever get to the point where the bra is just good enough the way it is? It's not like the bras today are that much better than bras from ten years ago.

So if you buy the Ipex, you will own the most advanced bra in the world. But none of it really matters, because it all comes down to the mechanics of the human touch; having someone who knows what they're doing. Maybe scientists could work more on that.

# Dog's Life

I heard about a dog committing suicide. But the dog was blind, so it was up for debate. One neighbour believed the dog knew what he was doing when he 'jumped' down the stairwell. Another neighbour swore it was an accident and he slipped. Given the dog was old, that was also a factor. 'They should have put that dog in therapy long ago,' said the man who insisted it was suicide.

In New York, even dogs have therapists. When my friend Meagan had a baby, her dog, Bud, was put on Prozac because he wouldn't eat or wag his tail. Finally, a dog I could relate to.

I've never wanted a dog before, but a dog that wants to lie in bed all day and doesn't care about going out? When I heard that I said: 'Bring him over.'

He arrived with two cases and his own bed. I was feeling a bond. The dog packs more than I do: special food, toys, his hairbrush and his pills.

Dogs in New York are so stressed that in the past two years three dog spas have opened in my neighbourhood. And a doggy gym for the ones who don't like walkies. I wonder if dogs size each other up in the gym. Would a mutt be considered less attractive then a poodle?

The popular dogs must make the antisocial dogs feel bad. I've always thought a dog's life was problem-free, but maybe it's just as rough for them as it is for the rest of us. I wonder if there's a stigma attached to being depressed when they're around their friends. It's not like they can hide it if they're not sniffing the way they used to.

Being a dog in New York is brutal. Chances are you live in a tiny apartment with very little light and a view of the lamppost, twenty floors down. If you belong to a single woman, your only job in life is to be upbeat and ready for a cuddle when your owner returns from a bad date or a friend's wedding. If your owner is a single straight man, you know you were purchased as a puppy as a way to pick up women. Now that you're full-grown and losing your cuteness, you've served your purpose and have anxiety attacks about being given away.

If your owners are a family, you've been ignored since the baby arrived or, if you're a playmate for the children, you're exhausted from being treated like a living stuffed animal. If your owner is a gay man, you're in a Burberry raincoat every time there's a drizzle, and if you belong to a model, your paws never touch the pavement.

The only dogs in New York that seem well adjusted are the ones jumping out of the Wagging Tail van on their way home from doggy day care in the country. That's because they don't have to go to the toilet on the sidewalk. Dogs always look embarrassed when they're pooping on the street. They have a look that says: 'Can I get a little privacy?'

There's a lot of pressure to be a cute dog these days so I don't think it's unreasonable a dog would commit suicide. Not being groomed enough, having a bark that's too loud, having to be cheerful because that's what your purpose is. Dogs worry about things too. I wouldn't want someone touching my nose all the time to see how I'm doing.

# Idioms and Hurricanes

I've discovered there is a name for taking all day to read the newspaper. It's called Parkinson's Law. It states: 'Work expands so as to fill the time available for its completion.' I looked into this. C. Northcote Parkinson was a 20th-century British historian. But will he be remembered for his writing? No. His legacy is the law he left behind, a truism of human nature. I want one.

How does someone get their own idiom? It's the next best thing to having a comet. Having a comet is more glamorous but I bet it requires you to be an astronomer. I looked into that too. You have to sight it first and claim it.

Halley's comet was said to have been sighted by Edmund Halley, who predicted it would return in 1758. Turns out that Halley had a rivalry with Isaac Newton. I suppose if your nemesis comes up with the three laws of motion, a good way to trump him is with a comet. What am I trumping my rivals with? Not much.

Having your name on an idiom might not be as exciting as having a comet, but it's better than a hurricane. I met someone recently called Katrina and thought it's a good thing she's a cheerful person. Imagine if it had been Hurricane Ariel that slammed into New Orleans? I'd be associated with tragedy and neglect even more than I am now.

But I was curious. Who was Murphy? His luck couldn't have been that bad if he managed to get his name attached to the most recognizable idiom of all time. Murphy's Law is so well known that there is a conflict over who has ownership of it. It supposedly originated in 1948 with Major Murphy, an engineer in the US Air Force. Entire books have been written about who was behind it. My favourite explanation is: 'A feckless Irish guy named Murphy, who was a builder.'

Soon I found that laws, adages, and idioms are all over the map. They are named after chemists, politicians, paediatricians; it's very random. For instance, who's ever heard of Littlewood's Law? This states that an individual can expect miracles to happen to them at the rate of about one a month. It was declared by Professor Littlewood – who died in 1977, presumably in the middle of the month before his miracle could save him.

'About one per month' is a bit vague. If you have five miracles in March and only one in April, you'll feel ripped off. Plus, what is considered a 'miracle' anyway? I think it's a miracle that I haven't got an amoeba from the tap water I use to clean my contact lenses. Is that my miracle per month?

You'd think, for a law to become part of the western world's lexicon, it would have to come from someone important, knowledgeable or old. But the most recent law on record I found was created in 1990. It's Godwin's Law and states: 'As an online discussion grows longer, the probability of a comparison involving Nazis or Hitler approaches one.' Mike Godwin is an American lawyer. This gave me hope. If Godwin's Law is out there, why not Leve's Law?

Choosing my law was tough. I was torn between four options.

Option 1: 'If someone's telling you not to get your hopes up, prepare for rejection.'

Option 2: 'If you suspect something's wrong, it is.'

Option 3: 'When you think you're right, you probably are.'

Option 4: 'All of your worst fears are true.'

# Discontinued

There are five words no woman wants to hear. Even worse than: 'Sorry, I don't love you' is: 'That item has been discontinued.' Love you can always find again. But once they stop making your favourite cookie? There's no solace.

Audrey is grieving. There was a heart-shaped vanilla cookie at a bakery in Manhattan that is no longer being made. She is in a bad place because it coincided with a break-up. Not only does the loss feel significant but it also feels personal. As though the one source of sweetness (literally) in her life has been taken away.

She confronted the bakery owner coming out of his kitchen. He explained that the supplier he used discontinued the vanilla and therefore, no, they wouldn't be making it. She kept repeating: 'But just find ANOTHER company' over and over, and finally he walked away.

It wasn't her proudest moment.

I know where she's coming from. I wrote to a company that suddenly stopped making a corn muffin I was addicted to – and urged them to 'rethink the situation'.

They wrote back explaining why it didn't make sense because demand was low. I guess my buying the same muffin seven days a week wasn't enough to keep the company afloat. Now when I think about it, I can't believe how many of those I ate – considering what was in them. I convinced myself the corn made them healthy.

It's like any relationship. When you're in it, you can't imagine life without it. But then, time passes and you think about how maybe it wasn't as good for you as you thought.

Wouldn't it be great if certain types of men could be discontinued? The ones you're addicted to and feel you can't live without – one day you wake up and discover they're no longer available.

There's something about an item being discontinued that is jarring. Suddenly, something you have come to rely on and seemingly need can disappear and it's entirely out of your control. Whereas when a relationship with an actual person ends, I just assume it's my fault.

Now when I really like something, panic sets in. Will I be able to replace it if it's lost? Do I buy in bulk? My friend Lori recently bought a pair of shoes that she thought were so great, she foresaw wearing them out and then not being able to buy another pair, so she bought two pairs. Turns out she hates wearing them.

Another friend is constantly searching on eBay for things that have been discontinued. Usually shoes. She had a pair of clogs she wore down to nubs, and kept looking for on eBay. She would find them in a size 36, not her size, and it was always the same guy trying to sell them. Back in 2004, he bought all the unsold size 36 pairs. I told her I wear size 36. She could buy them for me and then live vicariously through my shoe size. She declined that offer.

Once something is no longer available, it develops a mythic status. There are some Brits who mourn the loss of the Fuse chocolate bar with more emotion than they would a dead family member.

For Lori, it was a Diet Hawaiian Punch soda. 'It was in 1990,' she recalls, 'in San Diego, where my family was on vacation, in a town that must have been a test market for Hawaiian Punch. If I'd known it would be the last time I'd ever see that soda, I would have gone to every grocery store in San Diego and bought it all up and shipped it home. I think about that soda all the time.'

Maybe it comes down to wanting what we can't have. Whether it's a muffin or a man, is the fact that it's inaccessible what makes it desirable? If certain types of men were no longer available, I wonder how long it would take to get over it and move on. Then again, those types of men will never be discontinued because there will always be a market for them.

# eleven

## MY FATHER THE OPTIMIST

# Where He Went Wrong

My father who is eighty lives alone in Bali and I am an only child. If that's not a prescription for Xanax, I don't know what is.

What if something were to happen? What if he were to get bitten in the middle of the night by a spider or poisoned by a mango? Last time I visited him I never worked so hard at controlling my anxiety.

'What's there to worry about in Bali?' he asks. We're driving to lunch in his jeep as he says this so I decide to start there. The seatbelt is broken. Indonesian drivers – mostly on motorbikes – don't seem to pay attention to the dividing line in the road. The one thing I'm not concerned about is if the airbag works. Because there isn't one.

'I never saw anything wrong with this car,' he says. 'Until you arrived.'

My father loves the tropics and the jungle. Jews don't belong in the jungle. The other day I found out there is only one shrink on the entire island. That's got to be a lonely practice. I wanted to make an appointment just to see who's in the waiting room. Probably a New Yorker on holiday. But my father didn't know how to find him. He knew how to find the local masseuse though. I told him it wasn't the same.

Nevertheless, I got a massage. I was stressed out from all the bugs. The bugs in Bali are prehistoric. Cockroaches on steroids. And the spiders. There was a spider in his bedroom that looked like it should be in a cage. I'm still recovering.

The Balinese massage was strong. So strong, I worried I'd need a chiropractor afterwards. And then what if I couldn't find one? I'd be stuck. I wouldn't be able to walk. So I attempted to learn how to say, 'careful of my vertebrae' in Indonesian. It didn't work out, but luckily, my back was fine.

My father and I couldn't be more different. He wakes up every morning excited to start the day. I wake up every morning looking for bug bites I might have gotten while I was asleep. He is always up for new experiences. My idea of a new experience is the spicy tuna roll instead of the salmon cucumber roll.

What's more, he's extremely adventurous. And since the age of sixty-five it's only been getting worse. He has trekked in Nepal, hiked in China, climbed Mount Kilimanjaro, rafted on the Ayung River in Bali, run the NYC Marathon and explored the jungles of Kalimantan. He's also been on safari in Kenya, biking in Tuscany, trekking in the Dolomites and Zermatt. He also did a triathlon on Christmas Island and climbed Mount Gunung Agung, a volcano.

What have I done? Not much. I cleaned out my medicine cabinet. And thought about taking up boxing. For every adventure he's had, I have a phobia to match it.

Which leads me to wonder: where did he go wrong?

All of this optimism, it can't be healthy. His refusal to look on the dark side means every day is filled with one positive thought after another. It's difficult to be around that all the time.

But that's the problem with optimists. They fail to consider anyone else. Day in, day out, it's the same old litany of joyfulness. Life is a gift. Take every day as it comes. It's exhausting.

The more I think about it, the more I realize how important it is to worry. Without worrying about the worst-case scenario, how would we prepare for things like illness, tragedy and death?

When I first arrived in Bali I couldn't eat and didn't feel well. My heart was racing and I had to lie down. I asked my father to sit at the edge of the bed and I took his hand. For the next forty-five minutes I told him everything I was worried about. After that, I felt better. My heart rate had slowed and my appetite returned.

Naturally he's not a worrier. And he doesn't get bitten by mosquitoes either. But even more disturbing is that he believes the two are connected.

# Worrying Takes Practice

My father has never worried about things like what tomorrow will bring. His attitude has always been positive. Now that seems to be changing. Why? Because he's aging. All of a sudden, years of living in denial have caught up with him. He's wondering what it all means; what he'll do with the rest of his life.

Here's what he'll do. He'll suffer. Like the rest of us. Why should positive people have a free pass?

I feel bad that he didn't come to terms with this sooner, but it's his own fault. Laughing, smiling, looking forward to sunny days – what was he thinking? Sooner or later that kind of attitude just has to take its toll. Positive people are so busy smelling the roses they never take time to worry. Luckily, I can help. I've been thinking about illness, dying alone and being infirm, all my life. How to live with uncertainty and negativity is a skill that comes through years of practice. I've told him he can't expect to become a well-adjusted negative person overnight. I've worked pessimism into my schedule and got into the habit of embracing it. Like brushing my teeth. It's become second nature.

Take disease – I worry daily. I know it's only a matter of time before something gets me. The other day an elderly woman said to me: 'I never expected to get arthritis.' I was amazed. Who doesn't expect that? If my toe aches it's the very first thing I think of. Then I find out it's a bunion – but at least I've got the worst-case scenario out of the way.

Once you accept it's downhill, there's much less disappointment. I feel sorry for people who never think about old age, because when it hits it's devastating. The only way to avoid it is to wake up in the middle of the night with a blinding flash of realization and then have a fatal heart attack immediately. Then it's not a big deal. You've only worried about it for a few minutes.

I know when I'm older, I won't be in for a shock. I don't want to get out of bed most days, so I'm used to infirmity.

My father isn't used to infirmity. When he is down, he feels like there's something wrong. I tell him: nothing is wrong. It's healthy. It's just that he never let things bother him before, so now there's a cumulative effect. My advice is to let things bother him a little bit more every day. Once he's built up a tolerance, I can show him a routine. First, sweat the small stuff – such as who hasn't called back and what that means – early in the day, so that it doesn't keep you awake that night.

Then, just before you go to sleep, worry about the big stuff you can't do anything about. Things that are out of your control like when you'll die, where, of what, and how long it will take, etc.

After ruminating on this for an hour, it gets so exhausting you fall into a deep and satisfying slumber. The next thing you know, you wake up refreshed, ready to start a new day of worry. Most importantly, he has to remember this is what life is about. It happens to everyone. And five years from now, when he's three times as anxious, he's going to look back and wish he was as miserable as he is now.

# twelve

ROMANCE

# Trouble but Not Too Much Trouble

It was a Sunday afternoon and I didn't know what to do. I decided to go out. I walked around for a while, but then I got hungry and thought: 'I'll buy some marinated mushrooms.' After that I felt that I had a purpose.

I walked for a while until I came to the market that sells prepared food. I'd bought marinated mushrooms from them before, but as I approached the section where the mushrooms were on display I could see that for some reason, this time, there were onions mixed in. When did that start? People always have to change things. I don't like it.

I moved in closer for a better look and just at that moment the man behind the deli counter leant over and whispered to his colleague: 'She's trouble.'

'What did you say?' I asked. The two of them stared at me, not responding. But I didn't back down. 'I heard you tell him that I'm trouble,' I stated. 'Why would you say that?'

The deli man smiled and shook his head. 'No, I said, "Don't give her trouble."'

Now he was lying. 'That's not what you said: I heard you tell him I was trouble. Admit it.'

He stared at me. I stared at him. It was a standoff. Even worse, he had a smug look on his face because I'd just proved him right. Trouble.

What makes someone trouble anyway? Asking a question about onions? Or in my case, just looking like the type of person who would ask a question about onions. Making someone break their routine and go out of their way — that's trouble. But what is the alternative? Self-abnegation?

I should eat the onions and suffer the consequences of indigestion, heartburn and chronic bad breath just so I don't inconvenience the deli man? No, thank you.

I had an ex-boyfriend who, whenever I called, would answer the phone: 'Hello, Trouble!' It was a term of endearment and my being trouble was cute; something he enjoyed. Then one day, the enjoyment wore off. All of a sudden, trouble wasn't cute any more. It was annoying.

This led me to believe you can be trouble, just not too much trouble.

You have to know where the line is. It's the same with crazy. A male friend of mine has a new girlfriend and all he does is tell me how crazy she is. He also had a girlfriend before her who was 'seriously nuts'. I told him he always goes out with crazy girls and he said: 'I know.' When I probed him about why that is, he said it's because they're more exciting in bed – you never know what you're going to get, and there's a kind of 'I hate you' that makes the sex better.

Really? I've had loads of men tell me they hate me, but it's never made the sex better. Maybe that's because I'm the wrong kind of crazy. Men never know what they're going to get with me – but not in the way they'd hoped. I'm unpredictable, but in the way that makes them afraid to answer the phone.

I've always wanted to be the good crazy, but it hasn't panned out. The good crazy is ambivalent about plans and phone calls: she's too busy being off in her own crazy world to care. But my crazy is more of a challenge. It requires conversations about the relationship – how it's going, where it's going, and then when we discover it's going nowhere, what went wrong.

I guess my kind of crazy is just too much trouble.

# Roving Eyes

I went out with a man who would openly check out other women while he was with me. Sometimes, like a hypnotist, I'd have to snap my fingers directly in front of his face to redirect his gaze. Occasionally, he would even make a comment and he could never understand why it bothered me.

'You can check out other men, and I won't get upset,' he'd say. But that argument doesn't work. I'm not stimulated by a man's body. It's hard to glance at a man and say, 'Now he looks like someone who would be willing to listen to me for hours.'

When that argument failed, he offered reassurance that he wasn't leering, he was 'admiring', suggesting I should respect him for appraising beauty. That was good. His appreciation of a waitress with a nice ass should give him points.

Of course he assumed that if it bothered me, it must be because I felt insecure. No, I said, it bothered me because it meant he wasn't listening to what I was saying.

His response was: I am who I am. There's really nowhere to go after that, is there? The only thing that's worse than people who say 'I am what I am' is people who say 'I yam what I yam' while doing a Popeye imitation.

People who give anodyne answers are hard to converse with. Once you say 'I am what I am', you're just a sentence away from 'It is what it is'. Which in our case was: over.

I understand that when eyes rove, it doesn't mean feelings do. Which is why I don't feel an urge to check out men randomly. I stare at a man if I think I recognize him or I'm thinking about something and he happens to be in my line of vision. One time this happened and I smiled, but then he smiled back. Next thing I knew, he was heading over to talk to me. So unnecessary.

Lately I've been observing men who admire other women, and have discovered even the most seemingly mild-mannered men have strategies. My friend Tim told me he has a system. It's based on a risk-reward assessment.

The risk: how important is what the girlfriend might be saying at the time?

The reward: how attractive is the other woman?

'If what the girlfriend is saying is very important, I don't look. If she's silent, or what she is saying is not that important, then I look discreetly.'

So he'll fake paying attention rather than be honest? Without hesitation, he replies yes. The brain, he says, is big enough to work out how to check out the other woman without being caught, and pay attention sufficiently to stay engaged in the conversation. As for comments, he said he'd never make a comment out loud about another woman, but he'd feel free to comment about a handsome man. 'Admiring men aloud from time to time is a usefully misleading strategy when actually looking at women.'

Who knew there was so much technique involved? Now a whole new category of suspicion has opened up. If a man is looking for the waiter, I've assumed he wants the bill. But maybe he's using it as a ruse to check out the woman near the bar.

One thing I know is that if a man's going to look, you can't change him. So I make sure that he sits facing the wall. Unless it's a wall with a mirror.

# An All-Male Brothel

Recently I read that there are plans to open the first all-male brothel for women. If it happens, I predict a multitude of complications.

For instance, I can't sleep with someone without wanting to talk on the phone afterwards. Part of the deal would have to include at least an hour of talk time the following day so that we could do the post-mortem. During that call I'd be able to ask as many questions as I wanted. And this would have to be included in the price.

And what about separation anxiety? As soon as whoever I was with would tell me it was time to go my response would be: 'Already?' Then I'd want to know where he was going and who he was seeing next.

He'd say, 'It's none of your business,' and we'd get in a fight. It would end in tears. I'd want my money back.

If a male brothel is going to speak to the fantasy women have of men it would list their qualities as opposed to their measurements. 'Kind, generous, considerate.' Or: 'The type of man who would never break up with you by text message.'

I'm curious too what the line-up of men would be like. There would have to be a really old gentleman – for women who want a father figure. There would need to be a man with a great sense of humour. It wouldn't matter what he looked like as long as he made you laugh. Then there's the ambivalent man. Lots of women will pick this one because he's familiar and reminds them of someone they've dated. As soon as you pick him, he seems really excited to spend time with you. But a few minutes later he's not sure.

You go to the room anyway and hope the ambivalent man will show up. Maybe he will, maybe he won't. He'll keep you waiting, confused, and then just as you're ready to give up and move on, he appears – and you love it.

Audrey said if she went to a male brothel she'd want to have sex with an office worker. Not a high-powered businessman, but someone who works in the IT department. 'Someone who, under normal circumstances I'd probably never sleep with,' she said.

This would be an adventure because of the bizarre unlikeliness of the scenario. Like a man meeting a nurse who actually looks like Pamela Anderson.

Sophie says she would want someone who would talk only about all the things they were going to do in the future. Why? 'Because most of the men I know get skittish when I mention breakfast.'

In her fantasy the man she's with whispers in her ear: 'I want to buy us non-refundable tickets to Paris for New Year's Eve 2014.'

I'm not sure what my fantasy man would be. Probably a doctor. Not a dermatologist though – they never have good news. He'd be dressed in scrubs and have that fresh-out-of-surgery look. Only when I said I wanted him to examine me, it wouldn't be a euphemism.

I would spend the entire time quizzing him on things I think I might have – and even if he couldn't answer I'd still have an hour of his undivided and intimate attention. What could be more pleasurable?

Here's how the pricing would work. $1000 buys an hour of uninterrupted talk. $2000 gets you a really good foot massage while he listens to you ask his advice about an ex-boyfriend's new girlfriend.

The most popular man at the brothel? The cuddler. And runner up would be the one who wanted to stay in bed and watch *Tootsie*.

Finally, there's something else to consider. Brothels appeal to men because it's sex with no strings attached. But for women, that's not hard to find. If I announced I'd like to have sex with no expectations, demands or agenda – I'd have to turn men away. I certainly wouldn't have to traipse out to the middle of Nevada to find it. So while in theory a male brothel sounds plausible, in practice it wouldn't work.

It's sex with some follow through that women would pay for.

# With a Woman

Someone asked me if I'd ever thought about having a relationship with a woman. Of course I've thought about it. I've thought about how much easier it would be. But that's where the thought ends. Because I'm attracted to men, so it's like thinking about how wonderful it would be if I were taller.

But let's say that it was possible. If I were in a romantic relationship with a woman, I could call any time without fearing that I'd be accused of being too needy or clingy or desperate. Even if I called her at work to tell her I felt fat.

Here's a conversation I bet we could have. 'Hi, honey, I'm feeling bloated.' She'd say: 'Me too.' Then we'd say: 'Love you' and hang up. No big deal.

Can you imagine calling a boyfriend at work saying: 'Hi, honey, I have an ingrown hair'? He'd never want sex again.

I bet the fighting would be less stressful too. With every man I've been with, the argument unfolds like this: I'll get angry, he'll walk out, usually muttering something like: 'Who needs this?' Or he'll get angry, keep it to himself, and do the crossword. Then when I ask if he wants a cup of tea, the response will be: 'You're such a nag.'

But fighting with my girlfriend would be an extended therapy session. Women communicate. Fighting would be just like talking. Only louder.

Also, situations that generally get ignored by men would be taken seriously. Like having to go out with frizzy hair. My girlfriend and I would sit down and discuss the options. Pull it back? Put it up? Headband, hat or scarf? She would implicitly understand the psychological effects of the frizz and genuinely sympathize.

When I've mentioned having frizzy hair to a boyfriend, his reaction has been one of two things. Either he'll offer a compliment like, 'I like it frizzy,' thinking that as long as he finds me desirable, that's all the reassurance I should need. Or he'll snap, 'What do you want me to do about it?'

That's because men are problem-solvers. Especially in relationships. They have to resolve something the instant it's a problem. The alternative is to endure the complaint. 'What can I do to fix this?' is what they'll say, but the unspoken part of that sentence is: '. . . so I don't have to listen to you whinge about it.'

With a woman, I can see how the problem would be savoured. Nothing would have to be fixed, changed or improved. We could just go round in circles for as long as we wanted. There would be no fear of hearing what's wrong, either.

When my ex-boyfriend asked this, he dreaded the answer. I'd say: 'Nothing' and he'd say: 'Good.' My girlfriend would understand. Nothing means everything. Then we would get into a protracted discussion and dissect why life was such a mess. Joy.

The worrying part is what happens when we break up. I wouldn't be able to use gender as the reason. With a man, there's always the excuse: he's a man. That's why the relationship fell apart. But if I were with a woman? I would have to face the truth. Maybe it's me.

# Food Sex

I called a friend from a taxi and told her I'd just done an interview with someone famous over dinner. She immediately interrupted. 'Where'd you go?'

'Asia de Cuba,' I said.

'Ooh, I love that place,' she purred. 'What did you eat?'

I said I had the barbecued salmon. There was a mournful sigh. 'And?'

I paused. 'Mashed potatoes,' I said, 'with chunks of lobster.'

Suddenly I realized we were having phone food-sex. I knew she hadn't had carbs in a while, and talking about potatoes was driving her wild. I decided to take it to a really naughty level. I mentioned dessert.

'And then,' I said, lowering my voice, 'I had . . .' Her breathing got heavier.

'Go on,' she said. Slowly I uttered three words. Dulce de leche.

'With the coconut cake?' She was swooning. 'Did it have the icing?'

Before I could answer she cut me off. 'Stop,' she said. 'I've had enough.' It had become too much. Somehow it had gone from food sex to food porn.

There really should be a food-sex line. Women could talk to a stranger about food for as long as they wanted. And, unlike the women who operate phone-sex lines, these women would enjoy it. They'd ask, 'What are you eating?' or 'What are you chewing?' and it wouldn't sound gross. For an added charge they'd recite recipes.

Very few men like talking about food on the phone. If I say to a man, 'I've just ordered dinner,' chances are he'll say: 'Call me when you're done.' A woman, on the other hand, will always want to know where I ordered from and what I'm having. We'll go on for so long, my food will arrive, I'll pay the delivery man and we'll continue, giving a running commentary as if it's a sporting event.

'They didn't give me enough extra ginger. Oh, here it is. The yellowtail looks rubbery. No, hang on, let me taste it . . . okay, it's good.'

The only question a man ever really seems interested in asking over the phone is: 'What are you wearing?' I know when a man asks what I'm wearing, he's bored. But when a girl friend calls and asks what I'm wearing, it's another story. It's because she's genuinely curious. There are things to consider: hair, weather, function and so on.

Men think we talk about them all the time, but we don't. What's there to say? Unless there's a problem, there's nothing interesting to discuss. When it begins to sour, then there's a topic.

Food is a different story. Audrey will call me from the bakery to tell me about a new cupcake and I'll have to know the details. Is it red velvet or carrot? Is the frosting thick or thin? Sprinkles? What does it smell like? And the words 'You can't lick it off' are very exciting.

I know there are other things I could be talking about. Politics. Making poverty history. But a conversation about cupcakes is so satisfying. There must be more to life than this. But maybe not.

# If a Woman Fancies You

Recently there has been a proliferation of articles and lists offering men common-sense advice about women. The latest one: obvious ways a man can tell if a woman fancies him. The audience must be men who have lived in a cave. And the list is misleading. It's not stating the obvious. It's creating false hope and contributing to more confusion.

For instance, if she sits on the edge of her seat when talking to a man she's 'interested'. Not necessarily. I sit on the edge of my seat all the time. It usually means I'm about to get up to pee.

Here's another: if she drinks beer from a bottle (instead of Merlot from a glass), she wants you. If you're the sort of man who would interpret this as a signal, you shouldn't be dating.

And another: if she excuses herself to apply lip-gloss it means she's attracted and if she applies it in front of you – it's a bad sign. Really? I apply lip-gloss in front of men all the time. Then again, that one could be right.

Apparently you can tell she's interested in you if she doesn't answer her mobile phone while she's with you. This definitely is false. She might not be answering her phone because it's her boyfriend wondering where she is. Or her credit card company chasing her for the monthly payment. Or, in my case, I'm not answering because no-one is calling.

Also on the list: if she touches you, more than once. Again, not so convincing. I touch men on the arm if they're not listening because I want them to pay attention. Just because I want them to listen to me doesn't mean I want to date them. Doesn't this also depend on where she is touching?

I suppose there might be a few men out there who can't tell when a woman is into them, but I've never met any. In my experience, men always assume women are into them and if not, they assume she must be a lesbian.

# Howdy

Len Rosenberg called. Again. I have already told him, several times, we are not a match. I pick up the phone and he says, 'Howdy, Len Rosenberg here.' If you're going to woo someone, don't open with 'Howdy'. Especially if you're a Jewish man living in Manhattan. He then asks me what I'm doing this weekend and if I want to meet for Mexican food.

I tell him I can't and that I'm in London. 'I guess you never know when you call a mobile phone where you're calling,' he says. But he sounds like he doesn't believe me.

It's true I could be making it up, but since I've already let him know I'm not interested, the need to pretend to be out of the country seems unnecessary. Still, feeling like I have to prove to this person who I don't want anything to do with that I'm not lying, I point out the noise that he hears in the background is the Tube announcement and that I'm waiting on the platform.

He tells me to stay off the Tube.

Why? He knows I'm waiting for the Tube, about to get on it. What does telling me to stay off it do other than make him feel caring and make me feel anxious? Thank you, I say, intending to sound sarcastic but he takes it literally. 'No problem.'

He asks what I've been up to and I mention I was in Italy (which I had told him already before I left). 'Wow,' he says. 'You must have gained some weight over there eating all that pasta.'

Then I tell him I was in Israel too which he doesn't seem the slightest bit interested in. So I decide to lie and say I was in Gaza, just to see how he responds. I hear him chewing gum at the other end. After a few seconds he says: 'Cool.'

His other line rings. 'Listen,' he says abruptly, 'I'll let you go.'

He'll let *me* go. How did that happen? I was the one who was ready to go from the second I heard the word 'Howdy'.

# Blind Date

Blind dating is back. But so are cowboy boots; and just because they're back doesn't mean I plan to get a pair.

Recent predictions have stated that there will be an increase in terrorism, global warming and blind dating. The first two I can handle. But a blind date? Cooling the Earth is less daunting.

What's wrong with internet dating? Who wouldn't prefer to sit alone in a bathrobe, trawling through misleading photos, pretending to be charming in e-mails? I bet web dating got too easy. Maybe people are nostalgic for getting dressed up, having no idea what your date will look like, and being totally let down and forced to sit through dinner for two hours. Maybe they miss the old-fashioned disappointment.

I love when I hear celebrities met on a blind date. It probably helps ease the 'blind' part if you're famous and in movies and magazines.

The last time I went on a blind date — the only time I went on one — I was fixed up by someone I didn't know very well. This is the best way to go, because when the date's over you can end the acquaintanceship, horrified someone would think so little of you.

Having spoken on the phone first, we decided to meet at a restaurant in my neighbourhood. I asked how I'd recognize him; he described himself as 'white' and thought that was funny. It would be a long night.

I arrived first. An attractive man came in and seemed to be looking around. There was a twitch of hope. He started to walk towards me and I asked: 'Are you Bob?' Any answer other than yes (which includes 'I could be') would be reason to leave. It was a no. Just I was about to leave, the real Bob shows up. White Bob. How did I know it was him? We'd met before.

A few years earlier, we had spoken briefly at a party. My first thought upon seeing him again was, 'Oh you're that guy – the guy I was bored talking to,' which was only slightly better than his reaction, which was to stare blankly at me. Not only was there no recollection on his part, but he instantly knew exactly the wrong thing to say: 'We've met? Are you sure?'

At least we hadn't kissed. That's happened to me before too. Someone I made out with had no idea who I was. When we were introduced years later he looked right at me and said, 'Nice to meet you.'

And people wonder why I have low self-esteem.

Liza's recent blind date sounded more my type. At dinner she went to the loo. When she got back, he was rummaging around his pockets and produced a syringe and cotton swabs. 'Did they tell you I'm diabetic?' he asked. Then he lifted his shirt and injected insulin into his stomach. Liza lost her appetite. 'The rest of the evening, all we talked about was diseases,' she said. And that's bad?

# On Paper

I once went out with a millionaire. We met at Lori's party and she told me why 'on paper' he was a catch. Tall, good-looking, well read, actively seeking a girlfriend. It was impressive.

What am I like on paper? If Lori were to describe me she'd have to say: 'Talks on the phone, drinks coffee.' At least it's not misleading. I'm so much better on paper than I am in person – and on paper, I'm not that great.

The millionaire was the bachelor that everyone wanted to date. Some girls would see that as a challenge. Not me. I saw it as a reason to stay home, since chances were it wouldn't go anywhere. I was already exhausted from the effort of trumping every single woman and I hadn't even heard from him yet. But the following night he called and we went out to dinner. It was pleasant or, as I told Lori, it was less horrible than I had expected. 'Well, he really liked you,' she said. She sounded stunned.

So we went out again. He invited me over to his flat. I could see myself living there; just not with him. Here's why. Without knowing me, he had decided that I was wonderful and all of his friends would love me. And then, the more I tried to convince him he didn't know what he was talking about, the more charmed he became. It was a mess. The evening went on. He suggested a nightcap and I said I had to get up early the next morning for a meeting, a really early meeting, and added '6 a.m.' to emphasize.

'I love it,' he said, 'you're so driven.' I didn't say anything. He walked me to the kerb and asked for another date. Then he kissed me.

Lori told me later he thought it was a great kiss. I don't know how this happened. How could he think it was a great kiss? Wasn't he there? It was like kissing my arm.

'What's the problem?' she asked. She didn't understand. He'd brought dried figs and expensive apricots to her party. He was perfect. I thought it over. It wasn't the fact that he liked me: it was the fact that he'd idealized me and had no idea how wrong he was. I almost wanted to date him for a few months just to prove I was right. Six months down the line I could gloat. 'See? I'm not so wonderful now, am I?' That would show him.

# Now You Know

After all the endless lists about men that women should avoid, here's a list for the men. There are plenty of men who can't read the signals when there is a woman to keep away from. They fail to see the signs until it's too late.

So even though it's a long way from burning with desire to burning your house down, here are some clues for men so that there's no grey area.

1. She asks you what you're looking for in a relationship and when you reply you're not interested in a relationship she mentions your beautiful eyes.

2. You cancel plans because you're too tired and she asks: why? When you tell her you've been working hard, she asks: on what? You mention a female co-worker and she asks: who's she? After you explain everything she says: okay, get some rest.

3. After talking about your wife, she asks if the marriage was for a green card.

4. When you tell her you don't think you love her she replies, 'You do, you just don't know it yet.'

5. After sex when you get up to go home that night and don't

call the next day or e-mail or text, she decides you lost track of time and sleeps with you again.

6. You break the news you don't want children (knowing she does). She suggests getting a dog instead.

7. She prefaces most sentences with, 'I've never told this to anyone before . . .'

8. When you talk about yourself non-stop, she doesn't interrupt.

9. After two dates she's put you down as her 'in case of emergency' contact.

10. She volunteers to get an extra set of your keys made – for herself.

11. She's become friendly with your mother but you didn't introduce them.

12. She calls you and says: 'It's me' after having met once.

Often with these kinds of lists people question whether the list-maker is credible. But this kind of advice isn't run-of-the-mill and could only come from experience. Enough said.

# Friends with an Ex

Going out with an ex has its advantages. The best thing is the feeling of having made progress. It's not often I get to feel that. Even though not much has changed in my life since we broke up, the simple fact that we're no longer together is enough for me to feel I've grown.

I went out to dinner with someone I dated for a few months. When

I got to the restaurant, he was smiling. 'It's so good to see you,' he said. I could tell he meant it. If only he'd looked at me like that once when we had been going out, we might still be together.

When we were seeing each other, I couldn't ignore the sense that it wouldn't last. One night I asked him what he was looking for and he said, 'I'm not looking.'

Hard to live in the moment after that.

But now that there is nothing at stake, I can be the girlfriend he always wanted. For instance, at four o'clock in the afternoon he called to let me know he might need to cancel dinner that evening. 'That's fine,' I said. 'Just let me know either way.' He was amazed. 'Why couldn't you have been this easy-going when we were dating?'

Because now, I don't care. Why is that so hard to figure out? As long as I'm not emotionally invested, I don't feel rejection.

When I'm not counting on something, there's no disappointment if it doesn't happen. It doesn't matter when he cancels because I haven't planned for a week ahead of time what I'm going to wear. Also, I don't take the cancellation as a sign that I'm not a priority. I already know I'm not a priority. We established that five years ago when we broke up.

Not caring means there's no jealousy, either. He can talk to me about his sex life with his new girlfriend as much as he wants. He might as well be talking about the stock exchange. That's how little it matters. I can listen intently to what he's saying now, too, because instead of wondering what he really means, all I feel is relief that I no longer have to second-guess every word.

Spending time with an ex is easy because there's a familiarity without all the complications. Things don't have the significance they once had. For example, now when he says, 'You deserve to be with someone special,' it's sweet. When we were dating and he said this, it was depressing.

Also, everything that was the problem in the first place becomes affectionate nostalgia. Over dinner he was talking about his current girlfriend and he mentioned that what he liked about her was that she wasn't needy. 'That's the complete opposite of you!' he roared. 'You never knew when to let go!' Oh, the memories.

So now that I don't care any more I could relax. Which led me to a crucial realization. I'm so much better as an ex-girlfriend than I am as a girlfriend. From now on when I meet someone new, I should give them a heads-up and say: 'You'll hate me when I'm your girlfriend but after we break up, you'll fall in love.'

# First Date

I read about a young woman who went on a first date and was hit by a truck. That's got to be awkward.

If you're the guy in that situation, it's hard to know the appropriate time to say good night. Before the ambulance arrives? If you're polite you should at least accompany her to A&E and wait for the family.

If I were hit by a truck on a first date, the guy would be in over his head. Because when he asked who he should call I'd say: 'No-one. It's just you. And don't leave.' Then I'd grab his hand and whisper: 'Promise?'

But, given the guys I date, one of two things would happen. Either he'd let go of my hand, place it back on the trolley and say, 'Sorry, you're on your own,' or the instant that I was wheeled into surgery he'd bolt.

Of course, if he was the one hit, that would be ideal. Tragedy bonds people. I could see us walking along, laughing, making plans for the weekend and then suddenly – boom – I'm in the hospital with him, making plans for his physical therapy and he's thinking how lucky he was to have met me.

Plus, he might be in traction. I'd have his undivided attention. Unless he was medicated, in which case I'd have to repeat everything when the sedatives wore off just to make sure he heard me.

This got me thinking about some of the first dates I've been on in the past. All of the best ones were with guys that I never heard from again.

For instance, one of the most romantic first dates I've ever been on was with a man I'd met at a friend's party. He chose the restaurant, booked the table and was waiting for me when I arrived. Over dinner he said a lot of 'we should' and 'we could' and talked about the second date. Then he walked me home and we kissed outside my building. I said good night and floated upstairs. The next day he called. To let me know he was getting back together with his girlfriend.

The most vivid memories I have of first dates are the arguments. There was the capital-punishment guy: we got into a fight about the death penalty. And there was the Marquis de Sade guy: we fought over the pronunciation of Sade. There was also the Brooklyn guy. I told him I grew up in the city and he became defensive and said, 'Brooklyn is the city,' and I said, 'No, Manhattan is the city, Brooklyn is Brooklyn,' and then he called me a snob. That was fun.

I'm not sure what my perfect first date would be. Maybe a glass of seltzer with lime and a night researching medical symptoms on the internet? Then he could move some of my furniture around and fix my computer. Or, even better: I'd be invited over to a man's house for a game of Scrabble. We'd get into an argument over whether or not a word was legal and then we'd look it up together and he'd see I was right. I'd win. He'd take my hand, stare into my eyes and tell me how impressed he was. But then I'd screw it up. Because just when he was expecting it to turn romantic I'd wonder if he let me win on purpose to get the game over with.

# Saying I Love You

I find it exasperating when someone says 'I love you' too soon. I'm not sure I know the right amount of time before saying it, but I definitely know the wrong amount: twenty-four hours. If you've been dating for less than a day and you hear that, it's doomed. Even a month isn't long enough to have it count. People say 'I love you' when what they really mean is, 'Bye for now.'

It's one thing to say it and another to feel it. Audrey believes she has been in love more than three times this year. She falls in love more often than I've been to the gym. The other night she called to tell me her new boyfriend said *it* out loud. Those three words designed to make love official. When in doubt of your feelings, it's instant reassurance.

Who was the first person to say 'I love you'? In the ancient Hebrew of The Song of Songs, there is a lot of love but no 'I love you'. Then again, Solomon was said to have had 700 wives, so no wonder he got sick of saying it. Maybe the first person to say it was a mother to her child. A Jewish mother. As in: 'I plan to talk to you every day of your life to show you I love you.'

For those of us who are cautious about saying it there is the added dilemma of who says it first. In the past, when I've been the first one to say it, I've always regretted it. But once it's spoken, it's too late. It's not as though you can say 'Just kidding' afterwards. One time I said: 'I love you' and the response was: 'Thank you.' How awful is that?

Once 'I love you' is out there, it's a free-for-all. I've been in situations where it's taken up to a year before I'll say it but once I do, I can't stop. I'll say it a thousand times a day. 'I'm going to the post office now. I love you!'

I'm not sure why we feel so compelled to say it anyway. And why love and not some other, less vague emotion? If only 'I need you' had the same heart-warming connotations. When someone says 'I love you', it's romantic. When someone says 'I need you', it's scary.

There is an ancient Egyptian poem where the Egyptian pharaoh

Akhenaten established devotion to Ra, the sun god. But devotion is different. If I had a choice, I'd much rather hear someone say 'I am devoted to you' than 'I love you', because it sounds so much more committed. Maybe that's why it didn't catch on.

In Homer's *Iliad*, Paris says to Helen: 'Never I loved a woman as I love you now as the sweet desire holds me.' This might have been the first time the pronoun 'you' was used. But it still doesn't explain why expressing the desire out loud makes it legitimate.

If the first written 'I love you' is hard to pin down, the first time it was spoken out loud is impossible to identify. I'm guessing it happened when a cavewoman was carried over the threshold for the first time as opposed to being dragged by her hair.

I have found that the best time to say 'I love you' is right before hanging up the phone. I like to say 'love you' and hang up immediately. That way I avoid the awkward silence that usually follows. Or even worse, 'I like you too.'

# Engaged

All I want to do is sleep. I suspect it relates to my fundamental inability to embrace the fact that I'm alive. Only, I can't sleep. I lie awake ruminating about all the things in my life that could go wrong. Sometimes I'll have an idea. I'll think: 'This is a great idea. I should get up and write it down.' Then I decide that getting up, out of bed, will wake me up too much and it's not worth it. If it's really good, I'm confident I'll remember it. I never do.

One night I was lying there thinking about an e-mail I received from a reader. His name was Charles and he said he thought I should go on a date with him and then potentially we could get married, have some kids, and I'd be really happy. And he would too. Really? I doubt it. But I appreciated the brevity of the offer. It was very direct. Date, marriage, kids, happy.

He also offered to buy me an expensive egg-and-salad sandwich, which he would then make into a funny skull. I always say I want to meet interesting men. Does it get any more interesting than that?

I liked the e-mail because it made dating, marriage and kids sound like it was no big deal. You want Chinese – or you want to get married? It lessens the pressure. But even though I'm forty, I still don't feel the pressure – that's my problem.

I have a neighbour, Ellen, in her mid-forties, who lives down the hall from me with her cat. I can tell because she usually has white cat hair stuck to her clothing. She doesn't talk much, which I appreciate, so I don't know much about her, other than she lives alone and she's been in the building for ages.

The other day I returned home to find several police officers on my floor. Someone had reported hearing screams from Ellen's apartment. The police had gathered outside her front door, banging, pleading for her to open up. 'We know you're in there, Ellen. Please, just come to the door!'

Just then, my other neighbour – a single woman in her thirties – stepped out of the elevator. She asked me what was going on. I explained that Ellen was inside her apartment and the police thought she might be in danger or hurt and could even have been tied up and held captive by her boyfriend, who beat her. My neighbour looked shocked. Then she said: 'That one has a boyfriend?'

Everyone who wishes they were in a relationship, and isn't, can never get over how someone else is, but I'm not like that at all. Liza was telling me about someone she works with who got engaged recently, and she said it's as though that woman got the last ring on the planet. Turns out Ellen was fine: she and her boyfriend were fighting over the wedding plans.

I wish I could be more envious of other people's engagements, but I never am. If only I could believe that once that happened, it would all be okay, because then I'd be happy. Why am I not dreaming of living happily ever after?

Maybe it's because I can't sleep.

# Death Row Boyfriend

Going out with someone worse off than you has its advantages. When you complain about him to your friends, they'll nod their heads and tell you that you can do better and that you deserve more. It's very affirming.

So is contributing to dinner. I've gone out with men who don't earn a lot, and when I've offered to pay my own way or, better yet, the whole cheque, I immediately become the perfect woman. The problem is, I tend to assume my contribution gives me the right to be as miserable and difficult as I want.

If someone is worse off than me, it's not like he'll be the one to leave, right? Wrong. I was seeing someone who had no money, no job and no ambition. Three weeks into the relationship, he dumped me for some-one 'more upbeat'.

Finding the right mix of loser in a boyfriend is tricky. He has to have enough sex appeal and confidence to be desirable to me – just not to others. I went out with someone and the fatter he got, the happier I was. When we broke up, part of me was at least relieved that he was so big that no other woman would look at him. Of course right after we split, he started to train for the New York Marathon.

Recently I heard news of a former loser boyfriend. He fooled around behind my back, broke my heart and later married. The news was he'd divorced, lost his job, and was living in his brother's basement. Instead of thinking, 'Serves him right,' you know what I thought? I bet I have a shot now.

My dream would be to find someone who feels lucky to have me no matter what. Which is why someone on death row would be ideal. He'd always look forward to seeing me. And he'd stay on the phone as long as I wanted. What else would he have to do?

Plus he'd be all ears, because the only thing he could do while he listened to me talk was pace. He wouldn't be able to say: 'I have to go.' And who's he going to cheat with? The more I think about it, the more appealing it sounds. We'd never worry about our future together.

But then I'd get attached. I'd forget he was on death row and become very upset every time he brought it up. He'd tell me I was in denial and our visits would be spent arguing. To get back on his good side, I'd try hard to get him pardoned. It would give me a purpose in life.

Except as soon as he got out of prison, he'd be bolstered by his new-found freedom and dump me. Because he'd be free then he'd see me as the pathetic one for having stuck with him.

# Wife Material

What makes a woman wife material? I was thinking about this while watching the Republicans on TV. The moral superiority is too much. It's all about family. What about those of us who don't have one? Why not have a single woman for president — she'd be happy to work on weekends and holidays.

Recently my stove broke. Or, more precisely, recently I discovered the stove was broken. It might have been broken all summer, who knows. This might come as a shock to some people, but I don't cook.

I went online to see how much it would cost to replace it. The cheapest stove I could find was $450. That's a new pair of shoes.

I can't even remember the last time I used the stove. I think I opened the oven last year for a few hours when I was freezing and the heat didn't work. So I use it for heating. But I don't need a stove for that – I have a blow-dryer.

I called Sophie and asked if she thought I needed a stove. 'What for?' It was a good question. Boiling water? I could do that with an electric kettle. What else?

Boiling an egg? I could live without eggs. Boiling my contact lenses carrying case? Suddenly it occurred to me, I *do* cook. Boiling is cooking – there's a flame involved. I'm good at cooking things that need disinfecting.

Sophie has just moved into her new apartment. I asked if she has a stove. 'I think so,' she said. 'I haven't checked.'

We're from the same tribe. The Not-Made-For-Wife-Material tribe. She suggested I buy a microwave. A microwave? This isn't 1984. Who uses a microwave?

'I do,' she snapped. 'I like microwave popcorn.'

That settled it. I don't need a stove. If I ever need a hot meal I can always go over to her house. She's just moved around the corner and she's only been there a few days but she's already got a problem with mould. Also, her floors slope. And she's worried the building might not have heat. I asked why she was worried about that when it's still sort of summer. 'Because of spending time with you,' she said.

At least in my apartment I know that when something breaks or goes wrong my building superintendent will fix it. I have the best Super in Manhattan and unlike some of the other men who've been inside my apartment, he answers the phone when I call.

Incredibly, I've managed not to alienate him. Maybe because when he says he's busy, I believe him. And I know I can count on him in an emergency. It might be the healthiest relationship I've ever had. I appreciate him and he appreciates me. I'm quiet, I never have people over and I walk around in socks.

I might not be wife material but I make a great tenant. If you're a landlord – I'm a real catch.

There's a hierarchy of handymen in the building and I've discovered it's the difference between pitching an idea to a Hollywood development minion and the head of the studio. When my toilet gets clogged, I get to go right to the top. It's the one area of my life where I feel I've made it.

I'll never forget the time the flusher didn't work and after changing the flipper or the flapper or whatever it's called, my Super told me I might need a new toilet. Okay, a stove I can do without – but a toilet? That would be tough.

'How much do they cost?' I asked. He gave me a wink and whispered, 'I'll get you one from downstairs – it'll be from the building.'

Does it get any better? So I'm not married with kids. I might be single and stove-less but I still feel special.

# thirteen

## HEALTH CONCERNS

# Root Canal Alert

'See that?' my dentist said, pointing to a thick line in an X-ray of my tooth.

'You have the nerves of an eighteen-year-old.'

Of all the things I could wish to have of an eighteen-year-old, this would not be top of the list. How about the skin of an eighteen-year-old? Or the energy?

He then showed me another X-ray to show what a nerve in a normal tooth for someone my age should look like. The line was thin, barely visible. Mine looked like the Yangtze River. 'Usually, the nerves recede as we get older,' he said. 'But you're an unusual case.' Of course I am.

For two weeks preceding this visit, I had been unable to chew on the right side of my mouth. All my life I've taken chewing for granted. Chomping away on carrots and granola; never suspecting that I was just one almond away from a lifetime of mashed potatoes and soup.

And suddenly it happened. The electrifying tenderness of biting down and feeling as though I'd been shot. I went to the dentist. He told me not to worry. But he always says this, and I worried he wasn't taking me seriously.

I remembered all the times I'd moaned about a cavity; how could I have used up my sympathy on a filling?

He removed my old crown, placed a temporary on, and then came the dreaded words: root canal. In the past, when I've complained about something, my friends would roll their eyes and say: 'It's not root canal.' What would they say now? It was almost worth getting the root canal to find out.

I told everyone. The reaction from my British friends was very different to that of my American friends. Most of the Brits I spoke to regarded root canal as slightly more eventful than napping. It happened on a lunch break and was not a big deal.

My American friends spoke about root canal as though it was open-heart surgery. Audrey said hers took five hours, 'and there was blood everywhere'. On the positive side, she noted, my root canal couldn't possibly be worse. That was the comforting part. That I wouldn't emerge covered in blood.

I couldn't understand how in London a root canal could happen at lunch, but in New York it would take three visits. Everything in New York is more complicated. And it requires a specialist. Now I won't just have a dentist, I'll have an endodontist. Soon, I might need a periodontist for my gingivitis and an orthodontist for the teeth that have begun shifting.

I can celebrate with a bowl of stewed prunes.

Two weeks later, back at the dentist, it was a scene from *Marathon Man*. He decided to remove the crown without numbing me up so I could test the bite.

He leaned in with what looked like pliers in his hand. I screamed so loudly, I could hear his other patients instantly rescheduling their appointments. Worse than the pain was the anticipation that in the next second it would be twenty times more excruciating. The real test came when it was time to see if I could bite down without being in agony. I bit down on a piece of pineapple.

Slowly, at first. But I could chew. I was filled with joy. Seconds later I was filled with regret. Because, after all the discussions about root canal, in the end it turns out I wouldn't even need one. Which means next time, when it happens for real, no-one will believe me.

# Clean Bill of Health

The rate of corrosion is accelerating so rapidly it's hard to keep up. My dentist has told me I have to replace all my fillings.

'Thirty years of wear and tear takes it toll,' he said. Am I chewing that vigorously?

He said it was most likely from grinding my teeth at night due to stress. In that case, I'm surprised I have any left.

My body is like my apartment: suddenly everything has to be replaced. While some women my age are choosing between dinner at Nobu and a pair of Jimmy Choos, I'm choosing between a new refrigerator and a mouthful of decay.

A visit to the dermatologist is a buffet of maintenance choices. And I'm not even considering the cosmetic procedures – Botox and silicone are for the young at heart. I'm not looking to recapture my youth. My wish list is more basic. If there's a cancerous mole – get rid of it. Wrinkles are the least of my worries.

Liza pointed out the good news: at least I don't have a problem with my hip. She went to a doctor the other day because she had a limp. They were going over the 'what ifs' and the subject of hip replacement came up. She told him: 'Listen, it's hard enough to be a single girl with a cat – I really don't know if I can work a new hip.'

She was trying to think of a clever way to let him know she was single.

If it turns out that Liza does need a new hip, at least it will be because she earned it from doing something enjoyable – like dancing. If I ever need a new hip, it will be because I got out of bed the wrong way.

For my birthday this year, what I'd really like is a clean bill of health. Or even better, the health without the bill.

Since my eyesight is deteriorating, I might not have to worry about how I look. I had always assumed that if I slept all day it would lessen the strain on my eyes. Apparently not. Now I need a new prescription, and new contact lenses are expensive, so I'm wondering: do I really need to see everything?

I've seen enough.

# Future Husband

I bet if I married a doctor I'd be happy. When we're in bed at night, I could ask: 'Honey? Is this a melanoma?' He'd roll over, take a look and say: 'No, darling, it's a freckle,' and I'd smile. Reassured.

I would even tolerate dinner parties because I'd invite his friends over — urologists, gastroenterologists, cardiologists. I could interrogate them endlessly, learn about illnesses I never knew existed and then check myself for the symptoms.

Imagine going on holiday. The only thing better than travelling with a husband would be travelling with my own personal physician.

But in order to meet a doctor, I'd have to get sick more. I should start hanging out at hospitals. But that can get depressing. Plus, they'd be preoccupied — it's hard to seduce someone when they're in the middle of heart surgery. The best thing about marrying a doctor is being able to reach him at all times — because he always wears a pager. And when he wasn't with me, I wouldn't be suspicious; I'd know what he was doing — saving lives.

That's sexy. As long as he didn't spend too much time saving lives. Or working the night shift. Or holidays. Or weekends.

On my way back from vacation I met a man from Italy at Bangkok airport and we started talking. At first, the Italian charm was too much and I tried to politely exit the conversation. But then he told me he was a doctor. Immediately, he became more attractive. When he told me his speciality was infectious diseases, I wanted to marry him on the spot.

I began asking questions and he responded. This could be the one, I thought. He enjoys giving answers, and I enjoy getting answers: we're a perfect match.

But the more I asked, the more the charm evaporated. I asked if he thought it would hurt if I took an antihistamine even though I was taking Lemsip. I showed him the tablets that I'd just purchased — my ears were clogged and I was concerned that my eardrums would explode on the flight. He looked at the tablets. 'These instructions are written in Thai,' he said. 'I can't help you.'

What kind of doctor is that? I wasn't asking for a liver transplant. You can't tell someone you know how to dispense medical advice and then expect the conversation to move on. I asked one last question about bird flu, then switched to another section of the lounge.

The more I think about it, marrying an English doctor is the way to go.

English doctors are low-key. If I did have something fatal, he'd deliver the news with nonchalance. American doctors can be more alarmist. I wouldn't want my doctor-husband to freak out if I was dying. I'd need him to be calm — at least until I was gone. Then he could become hysterical.

I'm now holding out for a neurosurgeon. For our anniversary I'd get free CAT scans. Who says I'm not romantic?

# Colonoscopy

Here's a topic you don't hear a lot of Brits discussing: colonoscopies. When I found out that I needed one, very few of my British friends even knew what it was. When I explained — it's a procedure that examines the lining of the colon and small bowel, diagnoses gastrointestinal problems and screens for colorectal cancer — they seemed shocked that I would share such an intimate subject.

With my New York friends, it's another matter. If there's one thing New Yorkers love discussing it's medical procedures. Not only did they know what a colonoscopy was, but most of them had already had one. My friend Ben, who just turned forty, has one every year.

'You've never had a colonoscopy?' he said, sounding shocked. 'It's a piece of cake.' You'd think he was talking about getting his teeth cleaned.

At first, I didn't believe him. 'Did they give you a DVD of the procedure?' I asked. I didn't want to see it, I just wanted to make sure he wasn't lying.

'Are you kidding? I have a *Best Of*.'

Liza knew all about colonoscopies from the news anchor Katie Couric. When she was on *The Today Show* she had one on live TV after her husband died from colon cancer. Everything Liza knows about diseases she learned from celebrities. She knows about Parkinson's from Michael J. Fox, Alzheimer's from David Hyde Pierce, Hepatitis C from Pamela Anderson and so on. Who says celebrities don't make a difference?

Oprah devoted an entire show to digestion. Americans love discussing their bowel movements on national television.

At the Endoscopy Centre I went to for my procedure, the waiting room conversation was gripping. The middle-aged woman in the velour tracksuit next to me was loudly scolding her husband. 'Blood? And you couldn't leave work early?'

They put me out with an anaesthetic. Just before I went to sleep, the anaesthetist told me I had lovely blue eyes. I decided I should wear a hospital gown more often – it's a good colour for me. The last thing I remember saying before drifting off was, 'Are you married?'

He was.

When I woke up I was told the results and thankfully, my problem isn't serious. As the sedative wore off, I couldn't believe how rested I felt. It was the best sleep I've had in ages. I felt lighter too. No wonder all my friends get colonoscopies. It's better than a week at Canyon Ranch. Probably cheaper too.

# Sickness in London

When my British friend Colin moved to New York, the first thing I gave him was what every New Yorker needs: a list of good doctors and their phone numbers. My GP, dentist, chiropractor, ophthalmologist. The basics.

When I first got to London, what did people give me? An *A-to-Z*. How does that help? I'll be dead before I can locate a hospital on one of those maps.

Over the holidays, I came down with a very bad cold. Or, as I saw it, possible encephalitis. Not having a doctor of my own in London, I called my friend Simon and asked for the name of his. 'It's not like in America, where everyone has their own doctor,' he said. 'You have to be registered.'

I quickly discovered the choices in the UK were limited. I could walk to the local clinic and maybe they'd see me, but as it was Boxing Day, chances are they'd be closed. Or I could go to A&E. My symptoms were getting worse.

I called another friend and when she heard my voice she was concerned and suggested I see a doctor. 'Do you have one?' I asked. She didn't. How is that possible? All I wanted was the name of a doctor I could call and make an appointment, like a normal person.

'Call NHS Direct,' Simon said, 'and tell them you're not a UK citizen but that you need medical attention.'

I took matters into my own hands and went to the neighbourhood pharmacy. I told the pharmacist I was sick and he asked me questions like was my phlegm yellow or green? Finally, a productive conversation.

He offered a variety of products I'd never heard of, such as Betadine antiseptic iodine gargle for throat infections. There were lots of 'antis' on the label: anti-viral, anti-bacterial, anti-fungal. Loved it. Olbas Oil, an inhalant and decongestant, Robitussin cough syrup, Redoxon Double Action vitamin C with zinc, Tyrozets antibiotic throat lozenges. I got them all. I purchased more items in one visit than most Brits will buy their entire lives.

As soon as I got back to the flat, I opened up all my boxes of medications. My very own version of Christmas morning. I set them up and the kitchen resembled a chemistry lab. But there was a problem. On the directions for the Betadine gargle, it said to dilute 10ml with water. 10ml – how much is that? Not wanting to overdose, I called the pharmacist. 'I'm not sure if you remember me, but I just bought a bunch of items—'

He cut me off. 'I know who you are. The American.'

I asked if he thought it would be okay if I continued with the Lemsip even though it said on the label that people with Raynaud's disease should seek medical attention. 'Have you felt any loss of feeling in your fingertips?' he asked. I felt my fingertips. They weren't tingly – yet. 'What if it's a delayed reaction?' He said if that happened, to call a doctor. Then he hung up.

Later I called NHS Direct and they recommended a hospital walk-in service 2.9 miles away, in Hammersmith. I used the map. I had my ten minutes with a doctor and timed it. But it took an e-mail from my GP, Dr Samuelson, in New York to make me feel better. Not because of the medical advice: that was secondary. That he wrote back was enough.

## Sickness in America

Recently, I spent three hours on the phone. This may come as a surprise, but I didn't enjoy it. Why? Because I was trying to sort out my health insurance. It left me feeling so exasperated that now I almost hope I get hit by a bus and rushed to hospital – just so I know I'm getting my money's worth from my insurance.

Last month my friend Emily's insurance was cancelled. She found out when she went to the pharmacist who said her prescription wasn't authorized by her insurer.

'They don't handle individuals any more,' she said. As of 1 January they are only handling group plans. She picked up her medicine and had to pay $300 for it. Now she has to go on anti-anxiety pills because she's so stressed about paying for the anti-depressants.

Another friend is angry about the cost of her birth control. 'Like I even *want* to have sex now,' she snapped. 'It better be good.'

The pill costs $250 a month, which raises the question of who can afford to have sex? Between the cost of birth control and waxing, a boyfriend would have to be a luxury item. I suppose the good news is if I can hold out a few years, I'll be too old to get pregnant. That's the good news.

Health care in America is such a mess. You need a degree to wade through it. After months of examining different plans and affordable options, sifting through definitions of co-pays and deductibles, figuring out the difference between HMOs and PPOs, I signed up with a plan and paid extra for dentistry.

In no time at all I developed a toothache and discovered there is something in the small print of my policy called a deferral. I was told I have to wait a year before any major dental work is covered. At first I thought this meant I would need to wait a year before my claim could be reimbursed. But no, I had to wait a year before getting the actual work done. I can get a root canal done in January 2010.

It's gotten so absurd that those of us who have complaints about health coverage consider ourselves lucky – we can afford to complain. The other day I was speaking to a man who is getting divorced. When I asked what he'll miss most about his wife he said, 'Her health insurance package.'

Another thing. Hypochondriacs like me are a dying breed. I used to be able to see a chiropractor twice a month, not to mention specialists ranging from gastroenterologists to endocrinologists. But I've had to change my ways. Now, a visit to the GP is like a weekend in Paris. And

I'll use any opportunity to chat up a dentist. Which isn't easy.

I met a dentist at a party and I kept opening my mouth very wide while we were talking, thinking it was a good opportunity for a free check-up. It didn't work. He thought I was yawning and suggested I get some sleep. After that I tried smiling a lot. But he just assumed I was being friendly. He had no idea how unnatural that is for me – it used muscles I didn't know I had.

I decided to change my policy. After nearly an hour waiting on hold, the customer service lady from my health care provider asked, 'Are you a dependant?'

There was a pause. 'In what sense?' I asked.

She explained the coverage would cost less if I was married.

'Nope,' I said, 'Just me.' I told her I was dependent on people in other ways and reeled off a list of emotional hang-ups – but she told me none of that mattered. The one area of life where being dependent actually pays off, and I don't qualify.

# Bored Gynaecologist

The last time I saw my gynaecologist she seemed bored. Not with me necessarily, but with vaginas in general. Who can blame her? How many conversations about birth control can one person take?

I started seeing her at thirteen. That's several decades ago. I haven't had children or been pregnant. I'm still complaining to her about PMS and cramps. So, other than getting older and gaining weight, not much has actually changed.

There are many things about her I like. She's very clean and she smells of soap. She has a gentle manner I appreciate. This means when she tells me to get a mammogram, I don't immediately panic and assume I have breast cancer.

But like any long-term relationship, we've gotten into a pattern. Just as she picks up the speculum, she starts to small-talk. 'How's the writing?' That question makes me anxious under normal circumstances, let alone when I'm reclined and my feet are in stirrups.

I try to answer as best I can, but what I really want is to ask her to focus on what she's doing. I'm afraid she might miss something during our catch-up.

Sometimes I've wondered if I should see someone else. Shake it up a bit. There could be a disease that she's overlooked. Every question I ask is met with the same low-key response: 'Don't worry.' How is it she hasn't found anything wrong in twenty-five years?

I asked Liza if she would give me the name of her gynaecologist. 'Don't go to mine,' she said. 'She's the worst.' She is 'gyno to the stars'. If you're not Nicole Kidman, your vagina doesn't seem that important.

I moved on. I asked Audrey. 'Mine is cold and mean,' she said. 'You'd hate her.' I asked for the number. 'I'm not giving it to you.' She said: 'It's hard enough to get an appointment as it is.' Apparently, cold and mean is a hot commodity in gynaecology.

Emily, too, had issues with hers. 'She's depressing. Last time she told me my eggs were getting old.' That didn't seem so bad. At least she left with something to worry about.

I'd discovered a new phenomenon. Every woman I know is unhappy with her gynaecologist. And yet none of them switch. It's like people I know who have housekeepers. They complain about them but can't be bothered to find someone new. Why? Because they already know the routine. Who wants to explain it all again and get new keys made? That's reason enough to stick with someone for life.

A few months went by. I missed my old gynaecologist. When I went back, it was just like the old days. She seemed happy to see me again. I was happy to see her too. She was as enthusiastic as ever and things were back on track. I'm sure one day she'll find something wrong and I'll be grateful I stuck with her. Some things in life work out.

# The Tug Test

My hair was falling out. Not in clumps, or in patches; it was an overall shedding. I noticed that my ponytail, which has always been thick and coarse, had now become thin. Also, there was hair everywhere.

Several people directed me to Philip Kingsley, the wizard of hair. He is a trichologist and would be able to figure out what the problem was. I'd never been to a trichologist before. Who knew there was an 'ist' I'd missed?

I got to his clinic in Mayfair, entered his office and he shut the door. It's an old-fashioned study and he took his seat behind the desk. No phones, no disturbances – it was like having a private audience with the Pope.

A long time ago, he'd met my mother. He remembered she had lovely hair. Then he did the Tug Test. He ran his finger from my scalp to the end of my hair and examined what came out. Turns out, quite a lot.

He told me on a scale of one to four, I was a four. A four? That didn't sound so bad. Until he explained four was the worst. Which means on a scale of one to ten, I was really a twelve. But instead of being upset, I felt validated.

For months I had been telling anyone who would listen that my hair was coming out. No-one believed me. They wouldn't accept it, dismissed my concern and offered one diagnosis: stress. Then we'd get into a disagreement where I'd have to persuade them that my hair *is* thinning. I'd have to prove it. Not exactly an argument I want to win. But Philip Kingsley confirmed it.

He examined the follicles, the diameter, assessed the situation and studied my scalp. What was the problem? I'm pretty much doing everything wrong. This did not come as a surprise. I don't eat meals. I graze. Like a goat. Only a goat probably eats protein. Something I've been lacking. I have a lot of milk in my latte – does that count as protein? Apparently not.

'What do you eat for lunch?' he asked. Silence. I tried to remember what I had for lunch the last time I ate lunch.

'I only eat lunch when someone takes me,' I replied.

He could tell what that meant: not often. Dinner? Cereal. Or if I'm adventurous – soup. My food inventory was depressing. Prisoners have a better diet than I do. At least they eat regular meals – with company.

Me and my lonely bag of pistachio nuts hovering over the stove while I wait for my egg to boil. I might as well be bald.

After childbirth, sometimes women lose their hair. These women come to him, he told me, very upset, very worried – because as everyone knows, hair is intrinsically connected to our self-esteem. And losing it can be traumatic.

I pointed out that they have something to connect it to – giving birth. They might be losing their hair, but at least they have a child out of it. What do I have? I'm not exactly Elle McPherson. And my personality isn't all that winning. I don't have all that much to work with. If I'm ever going to meet someone, I need my hair.

He tried to reassure me that this was fixable. He could help. I would have my thick hair again, and it would grow back – it takes time.

'But how do you know?' I asked. He said he's never been wrong. I would get blood tests, and we'd find the problem.

Blood tests? I felt better when I heard that. For me, getting a blood test is more relaxing than getting a message. In the world of hair, Philip Kingsley is undefeated. He said he was certain the blood work would yield an explanation. But what if he can't help me? What if I'm that 1% whose hair loss is unexplainable?

There's a first time for everything.

# fourteen

NOT A FAN

# Trying New Things

Why would I try new things when I have more than enough trouble getting by with what I've got? Charlotte Brontë said: 'Better to try all things and to find all empty, than to try nothing and leave your life a blank.' Really? I'm not so sure.

Let's say I want to try murdering someone. I've thought about it. But then I spend the rest of my life in jail. Finding it empty will be the least of my troubles.

I do try some things. Teeth-whitening strips, for instance. I presumed they would change my life. If I had white teeth, I'd be happy. People with white teeth are always smiling. But they're just showing off. I'll never know because my teeth are still yellow. I certainly don't feel better just because I tried them.

Then there was the time I tried television comedy writing. Everyone said: 'Try it, try it — what have you got to lose?' Nothing. Except self-esteem and four years of my life.

People love to say: 'It's better to have tried and failed.' But what's wrong with not trying? Not trying means you don't have to live with the bitterness and frustration of failing. Even better, not trying means that trying is always in the future — something to look forward to.

When I think about all the things I haven't tried, I'm grateful. I haven't tried cocaine or bungee-jumping or firing an AK-47. I haven't tried kissing in a gondola or eating a blueberry bagel. I haven't tried having the mumps, either. Living life to the fullest, for me, means not having any health problems that would prevent me from staying home and not trying.

If I can see myself trying anything new, it would be a medical procedure that allows me to continue to keep things going just the way they are.

If you think about it, Charlotte Brontë was advocating disappointment. Now that's something I have tried. A blank life is more appealing. What is a blank life, anyway? A life without thinking.

My problem is I think too much. I even think about thinking too much. My ex-boyfriend would see me staring out the window and wonder what I was doing. I'd tell him I was thinking. Ten minutes later he'd ask the same thing, and again I'd give him the same response.

'Still thinking?' He looked at me like I was a different species.

Here's why trying nothing is a more preferable option. The more you have, the more you have to lose. Every time I've ever tried something new, it's led to nothing but trouble. I've learnt enough lessons from mistakes and shattered romances to last me until I'm 108. I have no room for any more lessons. I'm full. The learning curve has flat-lined.

# Adventure

Give me a five-star hotel that doesn't move and I'm thrilled. Boats — no matter how luxurious — are not my thing. So I was torn. I had a chance to meet my father in Rangoon and travel up the Irrawaddy River. But I get motion sickness. Boats, even when they're docked, move. And Burma sounds buggy.

Lots of mosquitoes. Which means shots and pills, which probably won't work.

Because even if the malaria-carrying mosquito has to buzz all the way over from Kenya, it will find me and make the bite.

My father couldn't believe his ears. Was I really going to pass this up? 'Where's your sense of adventure?' he asked.

But we all have different ideas of adventure. Passing up on an adventure is just a different way of being adventurous.

I've always thought of an adventure as a substitute for a catastrophe. When I'm on the verge of tears and need to stay calm, I'll remind myself that even if it is a tumour, it will be an adventure.

I met a friend for lunch once and asked him what exactly is the definition of adventure. He said it's to recognize when you are comfortable, and then to do something that makes you feel uncomfortable. Like switching shampoos?

Going by his definition, it's adventurous of me to get out of bed when I'm tired and wanting to sleep more. Staying fifteen minutes longer at a party than I would like is an adventure. So is engaging in a conversation about an article that appears in the Motoring section. Or wearing high heels and not taking a taxi.

I take risks all the time. Not to mention taking emotional risks. But no-one gives out medals for continuing to exist.

Recently I was about to order in miso soup from my regular restaurant. But for no reason, I decided to try somewhere new. I had no idea if it would be too watery, have too much tofu, or if the miso-to-seaweed ratio would be right. I didn't know if it would arrive hot or how long it would take. But despite this, I took a risk. The whole time I waited for my miso soup, I thought: 'There are people in the world who never try new things. I'm glad I'm not like that.' I felt good. Adventurous. The soup was disgusting.

I hadn't decided yet what to do about the holiday with my father. I called him and asked if I could do the trip, just without the boat. He was silent for a few seconds. Then he said: 'It's a cruise.'

# Local Travel

The last time I entered the UK I was told there was no room in my passport for the stamp. Not a single page left. I stared at the immigration lady in disbelief. How did this happen? I am someone who considers leaving the house for coffee an event. Yet I've become a person who requires extra pages?

Passports with extra pages belong to adventurous, fearless people. People who don't worry about getting typhoid or Japanese meningitis. Last week I took a bus across London and held my breath. Even more so when I realized I hadn't washed my hands after touching the pole. Later, I spoke about the experience the way mountaineers talk about K2.

If anyone who knows me were to open my passport they'd think it was a fake. Myanmar, Israel, Budapest, Italy, Ireland, Nairobi, France? Somewhere between showering and sleeping, I've been around.

Not that you'd know. When I think about the things I've done and the places I've been, I feel it should show more. There are people who I've met and just by looking at them I can tell – they've seen the world. When someone meets me, they seem shocked to discover that I've been anywhere other than to the doctor.

I've always travelled a lot. When I was five I began going from New York to Bangkok every year to visit my father. Every summer I'd be put on the plane with a laminate ID. Like Paddington Bear. Travelling halfway around the world didn't seem a big deal then. Now there are times when I have to prepare myself: You can handle going to a party in Brooklyn.

People say travel means leaving your comfort zone but I leave my comfort zone every time I go for a coffee at an unknown café. I let them know I prefer skim milk even though no-one pays attention. I'll say it's the cholesterol and that I'm not allowed to drink whole milk on doctor's orders. It sounds more official.

What are the chances that someone who I don't know and who I'll never see again is going to care about my cholesterol? It doesn't matter if I leave my house to go around the corner for a coffee, or around the world – it will always be a gamble.

Simon, who lives in east London, has given up on my ever going to his house. He left a message telling me that a friend had a car and would be willing to give me a lift. I told him I'm not a fan of road trips.

# Holiday

## The Odyssey Begins

Liza and I were on holiday in Italy for two weeks. Everywhere we went, she was smiling at men and calling out 'Ciao!' I, on the other hand, was calling out nothing. For Liza, a greeting is an opportunity to make new friends. For me, it's yet more people I'll have to avoid.

The first night, two men were flirting with us. In Italy, that means they said hello.

For no reason, she invited Giorgio and Giovanni along to dinner. But before we got to the restaurant, we were granted a tour of Giovanni's apartment, located above the local gelato store. Everything Liza found charming, I found worrying.

For instance: the ninety-year-old mother watching an Italian game show in the dark? Liza's reaction: how sweet he takes care of his family. My reaction: a thirty-eight-year-old man should not be sleeping six feet away from his mother. The photo collage of celebrities who have visited his gelato shop? Her reaction: impressed he met Kelsey Grammar. My reaction: time to go. The shelf of stuffed animals. Liza's reaction: he's a child at heart. My reaction: Liza's the least judgmental person on the planet.

At dinner, Giovanni discovered I was a writer and immediately volunteered to call his good friend, Gore Vidal. Seconds later, he hands me his mobile.

'It's Gore,' he says. 'Talk.'

Things were looking up. At first, he sounded confused. And a little scared. 'How do you know Giovanni?' he asked. I told him it was a long story.

Because the reception was so bad, I couldn't hear a word of what he was saying. I'm on the phone with Gore Vidal and all I can contribute is: 'Can you say that again?' I walked around trying to locate a better signal. Then the line went dead.

At the end of the evening, Liza is huddled discussing the finer points of gelato, and I'm left alone, having put the phone down on Gore Vidal.

## Week two

In an effort to communicate better with our neighbours, Liza learned to say: '*Lieto di conoscerla*' (Glad to meet you). I learned: '*Mi lasci in pace*' (Leave me alone).

She sits in the town square, chatting away and making new friends while I have mapped out the back roads to get to the places I need. The pharmacy, the bank machine, and anywhere that has shade.

In the evening, we went to a seaside restaurant. I ask the waiter in broken Italian, what kind of pasta the blonde woman at the next table is eating.

He stares at me for a second and replies in English: 'Luigi is Mama's nephew.'

A few minutes later, a loud family from 'the U.S. of A.' is seated behind us. The father enquires if they have Manischewitz, a kosher wine that tastes like grape juice. This is wine found at Bar Mitzvahs; not Italian seaside cucinas.

Throughout dinner, Liza is receiving sexy text messages from her new love, Giovanni. One of them reads: 'A sweet kiss for a special woman.' If this came from a guy in New York, she'd be rolling her eyes. But because we're in Italy, she's swooning and says he's a poet.

We rush back so that she can meet him at his gelato store in the square. He has not yet delivered on his promise for me to meet his 'good friend' Gore Vidal, but I'm told he comes to the square at night for sambuca, so I wait.

They go upstairs to his apartment while I sit downstairs, eating gelato, waiting for Gore Vidal. Three gelatos later, I call it a night.

I decide to develop a crush of my own. The local bus driver looks like Antonio Banderas but in order for me to see him, it means taking the bus along the winding roads of the Amalfi coast. Not the best for someone who gets car-sickness.

## Almost over

The Italian holiday was winding down and we were both having separation anxiety. Liza was preparing to say goodbye to her Italian boyfriend, and I was preparing to say goodbye to Roberto, the local pharmacist.

I developed a mysterious allergic reaction to a mosquito bite on my eyelid. Roberto told me he'd never seen anything like it. People say that to me all the time.

I think it was the only available place left to bite. I went to bed with socks on my hands and feet, long sleeves and pyjamas, cotton in my ears, and a scarf wrapped around my neck. Women in burkhas are showing more skin.

Everyone tells me it's because my blood is sweet. This makes me nervous.

I can't do anything to alter the taste of my blood. Liza, meanwhile, slept naked with perfume on. She wasn't bitten once.

Maybe I'm just not a holiday person. There are people who enjoy being out of their routine. Liza went to Positano, Scala, Capri and Vietri to buy pottery. I went to the post-office, the Internet café, the tourism office to collect a fax and the tobacco shop to buy more calling cards.

Essentially, I've recreated the same life I have in New York. The only difference is in Italy I'm covered in hydrocortisone.

I guess it's because holidays are about doing nothing and I always feel like I'm doing nothing. So is it that I feel that I don't deserve a holiday? I'm much too conscious of having to get back to all the nothing I'm not doing at home.

Every day of my life I do things to stave off the feeing of not doing enough and so when a holiday arrives, it's like sleeping naked: some people can endure it without a care in the world, whereas others like me, in spite of precautions, get bitten on the eyelid.

# My New Routine

I'm not a fan of change. Even if something isn't working I can probably live with it. Especially when it comes to relationships. No need for uncharted territory when there's a safe and familiar environment of misery. Why change when you can complain?

But recently, I have made one big change in my life. After years of searching for something to fill my days with meaning, finally, I've found change that I can believe in. Washing my hair every morning and eating three meals a day.

I know that might sound trivial but it's given me something to look forward to. Every day I wake up and there's a reason to get out of bed. Breakfast.

It turns out the hair loss problem I've been experiencing was due to not having enough protein. In order for things to improve I was instructed to keep my scalp clean and eat regular meals. Some people have religion, I have eggs and shampoo.

And like any devoted fanatic, I can't stop proselytizing. I'll talk to anyone who will listen about the benefits of breakfast. At least two or three times a day I find myself talking about eggs. I've always thought the only thing worse than being on a diet is listening to someone talk about their diet. But not when you're the person talking.

The subject never gets old. If only I was as enthusiastic about my career as I am about having breakfast, I'd be able to afford a cook.

In fact, friends tell me they haven't seen me this excited – ever. It's like I'm a Born Again. Scrambled eggs are my Jesus.

But it doesn't end there. Because not only am I eating three meals a day, I'm showering too, now that I have to wash my hair. Every day. Just like a normal person. Who knew being normal would be so exhausting. How do people do it?

I don't have a choice. I have to stick to this regimen and it takes time. There's thinking of what to eat, buying the food, making the food, eating and then cleaning up. Showering and washing my hair (two hours) then following up with the tonics and creams afterwards. There are also the vitamins – three different kinds twice a day – and the tincture at night. I'll stand in front of the mirror over the bathroom sink and instead of running through everything that's going wrong in my life, I'll focus on making sure the tincture doesn't get in my eyes when I put the dropper on my scalp. There's a big warning label on the bottle. It's very meditative.

The night-time routine eases my mind so that by the time I get into bed I'm relaxed. Then nodding off, I think about breakfast. I wake up the next day and it starts all over again. My days have structure.

Maybe it's blind faith. But isn't all religion? I have a vested interest in my spiritual wellbeing. My devotion just happens to be to taking a shower.

Here's the problem. It can't last. Three meals a day for the rest of my life? That's a lot of work.

At the moment, it's new. But what happens when the novelty wears off? Lunch will begin to feel like a grind. I can see it now – I skip one meal and promise myself it won't happen again. Then I wake up late and think: it's too cold for a shower; I can get away with just brushing my teeth. I won't go to jail. The next thing I know – I'm off eggs and having a double espresso and a pack of M&M's.

For now, I'm taking it one day at a time. I can't commit to a lifetime of change. I can't even commit to buying more than one roll of toilet paper. Besides, how do I know all this change will be for the better?

# Star Wars

There are two things in the world that everyone has done: seen *Star Wars* and had their ears pierced. I've done neither. But I've been opening my mind to new experiences. Maybe it's because I'll be dead one day.

Everyone seemed to be very excited about the new *Star Wars* movie, *Revenge of the Sith*, when it came out. People slept on the street two days before it opened just to be the first ones in to see it. I've never felt compelled to sleep on the street for anything. I wish I had it in me to feel that kind of devotion.

It might be hard to believe, but I've never seen *any* of the *Star Wars* movies. I'm a couple of decades behind. When I tell people this, they think I'm lying. '*You've never seen* Star Wars?'

Why would I lie about that? I didn't see it when it first came out and then as the years went by, I enjoyed the fact that having never seen any of the *Star Wars* films made me unique. Sometimes I'd worry: that's what makes me unique? But it never motivated me to do anything about it.

It was time. I went with my friend Simon who was elated to be seeing the film, just not with me. I had a million questions. Starting with: what's a Sith?

He told me to pay attention to the opening. What I needed was background. For instance, history on Luke Skywalker. He leaned over and began to whisper a lengthy explanation but as soon as I heard the word 'droid' I stopped listening.

That's when I decided I'm not a science fiction person. It's good to know these things.

The movie began. I looked around. I was the only one not watching the screen. All around me people were so engrossed, so enraptured; they couldn't have been happier. What was I missing?

It was like watching the galaxy screen saver on my computer. And the best part about that is, it's silent and I'm at home. After a while, I figured it out. The dark side, the Jedi Knights . . . it was escapism. I probably would have enjoyed it if I'd been escaping.

Just then, I felt bad for all the people who were enjoying it so much. Because that was it. The last one in the series. When it ended, what would they have to look forward to? What would they sleep on the street for now?

I sat there thinking about how it's important not to care about things too much. If I was the type of person who enjoyed *Star Wars* movies, I would have been very distraught.

## Making It in Los Angeles

I was in Los Angeles recently. I called Dan the Hollywood producer to see what was going on with the television show idea. He replied: 'It's always good to date a lot of people before getting married.'

As I suspected: nothing.

I told him I had to go to the computer store, and he said he'd give me a lift. Five minutes later, his silver SUV pulled up. When I asked what kind of car it was, he said: 'Porsche Cayenne Turbo.' From that point on, if I called it a Porsche, he'd instantly add 'Turbo', as if I'd left off a part of his name.

One of his hands was on the steering wheel, the other held his paper-thin BlackBerry. 'New?' I asked. He gave me a look and said: '8300.' I assumed that was the BlackBerry equivalent of Turbo.

I told him the reason I was going to the computer store was because I was upgrading my Apple software to something called Tiger.

'I'm the tiger,' Dan said, definitively. 'In Chinese astrology. I'm the tiger.'

Within seconds he had Evan, his PA, on speaker phone. 'Evan, what do you know about Chinese astrology?' Evan paused. 'Uh, nothing. But I can ask Ling-Ling, the Chinese intern.'

Ling-Ling, Dan told me, was named after a panda. He asked Evan to Google my birthday to see what animal I was. 'She's a monkey,' he said. Dan asked him to read out what that meant, but something was wrong. The traits were way too positive.

'Are you sure you have the right birth date – January sixty-eight?' I could hear his fingers tapping keys. Then he said: 'Oh, sorry, you're not a monkey. You're a sheep.'

The Apple store is at the Grove – an outdoor mall that is modelled after a European village square but feels more like Main Street, Disneyland. Dan dropped me off in front of a fake-brick path that led to the fake outdoors. Suddenly, out of speakers in the floor, came Dean Martin's voice: 'When the moon hits the sky like a big pizza pie – that's amore!'

Every fifteen minutes there's a new song that co-ordinates with fountains synched to move to the music. There are four songs on the loop and I knew I'd been there for forty-five minutes because I'd heard 'That's Amore', 'America the Beautiful', and 'Celebration'. People strolled around eating frozen yoghurt, carrying their shopping bags; a woman turned to her husband and said: 'It's like we're at the Bellagio in Vegas.' She sounded so excited.

Everyone in LA seems filled with hope. I left the Grove with my Tiger upgrade and walked to Dan's office half a mile away. We ate lunch nearby, since the Turbo was getting washed. It had a spot of mud on the left tyre. Over lunch Dan said he had someone he wanted me to meet for the show, but made sure to downplay it so I didn't assume anything would happen. He explained the situation and said he wanted to 'manage my expectations'.

What's to manage? They're pretty low.

Afterwards he offered to drive me back. 'Listen,' he said, 'chances are you're not going to make any money from this. This business is rough. I just don't want you to be disappointed.' I said it's okay, I'm used to it. So with him willing to take a risk and me being prepared to lose, we continued driving.

I was the sheep in a tiger's Turbo.

# fifteen

## AND ANOTHER THING

# Positive Spin

The weather in London is perfect: overcast and rainy. There's no expectation to be in a good mood. I think I've got a special case of SAD (Seasonal Affective Disorder). When the weather's nice I panic.

Instead of feeling on top of the world, I want to go back to bed. There are too many options. When it's raining, nobody expects you to go out or pays attention to what you're wearing. There's no pressure to be cheerful.

I wouldn't mind living someplace where the sun never comes out and people sleep all the time. I'd be the most productive person in town. The only problem is, lately I've been having insomnia. So there I'd be, wide awake, with nobody to talk to and nothing to do. Sounds a lot like my life now. Except I'd have frostbite. And I'm not a night person, either. Not having bright daylight is one thing; not having daylight at all is another. I need to know when the day's over: otherwise how will I know it's been wasted?

Recently I read that the Met Office wants to make the forecast more upbeat. Does everything need a positive spin? If people want their weather happy they should move to Disney World. Now, instead of it being 'chilly in areas', it will be 'warm in most'. Instead of 'occasional showers', it will be 'mainly dry'.

Why not just say 'mainly not gloomy'. When forecasters say 'mostly dry' we'll expect it to be, and then when it rains we won't have an umbrella, so on top of being emotionally unprepared we'll have ruined our favourite boots. It's far more practical to be negative. Forecast storms every day; if it's sunny it's a pleasant surprise.

Spinning things in a positive way seems to be the way things are going. I should start telling people that I'm not a negative person, I'm just mainly not seeing the positive. And instead of worrying about dying alone in my apartment, which has a 'great lobby', I'll expect to die surrounded by a bunch of wonderful people I never knew existed.

## Not Talking Pays Off

There are times in life when good things happen unexpectedly. Recently I experienced one of those times. I arrived at the Virgin check-in and was told that my ticket had not been processed properly, so I was directed to the ticketing desk. This is the third time in a row this has happened, and usually I complain. But this time I was silent. I couldn't bear to explain yet again what the problem was. Also, I was preoccupied because I thought I'd left my wallet in the taxi. And I had a sore throat.

'Thank you for being so patient,' the ticketing woman said. Then she whispered: 'I've upgraded you to upper class on your return.'

By accident she had mistaken my silence as civility.

It got me thinking about how my life would be if I was quiet more often.

There are so many perks I'm missing out on. For instance, in a relationship. What's the reward for staying silent and being patient? Not breaking up.

All I do is ask questions. Especially questions that provoke an internal search. 'What are you thinking?' and 'What are you feeling?' These are questions no-one enjoys. Especially after sex. Or when the TV is on. Or if the sun is out.

Questions that attempt to get information in general are never welcome either. 'Where were you?' 'Who were you talking to?' Worse than not answering is getting an answer along the lines of

'You don't want to know.' That just leads to a complaint.

A complaint is never in isolation. It's engulfed by wanting to know more and what can be done to fix it. And there lies the biggest question of all: would I be happier if I never complained? Probably not. If I had upgrades every day, then they wouldn't be special.

# Rubbish

I'm obsessed with the rubbish collections in London. I guess if you can't stop terrorism the next best thing is a crackdown on rubbish. A twenty-six-year-old bus driver was fined £225 for overfilling his bin by four inches. So now he has a criminal record. That same week I received a letter from Environmental Services.

'Dear Resident,' it began, 'Documentation with your name and address was found inside a sealed white bag under a "No Dumping" sign. A photocopy of the document is enclosed.'

Someone had gone through my trash? That was creepy. The letter went on to say that rubbish on the street on unauthorized days is an offence and carries a maximum fine of £2500 on conviction. Or, I could accept a fixed penalty notice of £80.

I contacted the person who owns the flat I was staying in. She was delighted. 'I'm so happy they're looking after the neighbourhood,' she said. 'People have been putting their rubbish out on Sundays and it's not right.'

Not exactly the response I had in mind. But if people had been putting their rubbish out on Sundays – why was I being targeted? I put the rubbish out on Tuesday, the day of pick-up.

'Yes, but it was a few hours early,' the gentleman on the phone said. I was aware that I was asking a lot of questions, so I mentioned I was a journalist. He replied, 'I know.'

How? He had gone through the trash. I was informed that once it's on the street it's in the public domain. I began to panic. What else had I thrown out? Account numbers, private information – old notebooks with bad writing.

How much does one's trash say about who you are? My trash has hypochondriac written all over it. It's filled with empty bottles and boxes of preventive medicines. Echinacea tincture and zinc tablets. It's got aluminium packets where all the Strepsils have been punched out and discarded bottles of First Defence. It's also got loads of doctors' receipts and printed out pages of illnesses I'd Googled.

There would also be empty packages for figs, prunes, dates, nuts and berries. My trash isn't very exciting. Unless you're a raccoon.

I asked Simon what was in his trash.

'Microwave meal containers, empty beer bottles and cigarette butts. What does that say?'

It says depressing. My neighbour, who juices everything, wouldn't mind someone else going through her trash. It's all compost, she announced proudly. 'My trash says I have a smallholding farm on the second floor.'

The following day I had to meet with the Cleansing and Enforcement Officer in person so that I could be legally cautioned and advised of my rights. We couldn't do it over the phone; he had to hand me the penalty notice in person.

I arrived at the offices at the appointed time. I was offered a cup of tea. Then we sat down in a room and he presented me with some of the items I'd thrown away.

'Would you like these back?' he asked.

Why would I want them back if I'd thrown them away?

After twenty minutes I was handed a ticket and told that was the end of it. That weekend, from my window I could see bags of rubbish piled up under the 'No Dumping' sign and it didn't seem fair.

The officer had supplied his mobile number on the initial letter. I called and left him a message alerting him to the rubbish situation and how it was four whole days before pick-up. He never called back. Now I'm an expert on the rubbish collections. It's three days a week but two of the pick-ups are consecutive: Tuesday and Wednesday. Why not spread it out? I've become way too invested.

# Low vs High Maintenance

There are two types of women in the world: low-maintenance or high. The low-maintenance women never complain. They will say, even though their spleen is bleeding, everything's fine. These women tend to have boyfriends or husbands who adore them.

They answer the phone with, 'Hello, darling,' and hang up without a fight even if the conversation lasts under a minute. They never shout, 'Wait, I'm not done!' And if someone asks how they are, they're always wonderful.

If someone asks how I am, I say: 'How long have you got?'

The low-maintenance women are afraid of being considered difficult. But being difficult isn't limited to personality alone. For instance, the men in Liza's life give her a hard time when she has to get waxed or have her hair coloured or cut, they complain about how much time and money she spends and how she's 'high maintenance'. But do they really want to be dating a girl with hairy armpits and split ends? Of course not.

Low-maintenance women are always quick to say we 'don't need all that stuff', but every woman needs that stuff. Some do it secretly and are just too proud to admit it. I'm very low-maintenance when it comes to grooming. Going to a hair salon requires a commitment to leaving the apartment.

The only appointments I keep are with doctors. I like my maintenance to involve a blood test or an X-ray. I wish I could be the sort of person who doesn't complain but that's not going to happen. I'm low-maintenance in all the wrong areas. A low-maintenance groomer with a high-maintenance personality is not considered a catch.

## Homemade Gifts

Does everything these days have to be homemade? The credit crunch has made people frugal – and industrious. Now it seems there's a backlash against anything that is store-bought.

I went to a party and took the hostess some chocolate brownies. She put them on a plate and passed them around. People seemed happy. Right up until someone asked: 'Did you make these?'

It's no longer enough to bring something, I have to have made it too? I said no, I didn't make the brownies and in an effort to redeem myself pointed out they were Fair Trade. No-one cared. The guests continued chewing. Disappointment hung in the air. A woman leaned in and gave me a sympathetic pat on the knee.

Instead of making money, people are making cookies. My friend Audrey was excited about a new man she's met. She loves that he bakes and declared: 'He's very self-sufficient.' Why? Because he knows how to make a peanut butter cookie?

Then I ran into a friend on the street and made a comment that I loved her hat. She took it off and handed it to me. 'It's yours,' she smiled. 'I made it.'

It was like a scene out of *Fiddler on the Roof*. Only instead of a Russian shtetl, we were in Notting Hill. I know times are tough but is this what it has come to? People are making things and giving them away. No wonder stores are closing.

Generally speaking, I'm not good at making things. I've tried in the past but it hasn't worked out. If you're going to make something it should be something to wear or something to eat. I discovered no-one really wants a poem for their birthday.

There are people who are not only good at making things but actually enjoy it. Liza, for instance, will always make a homemade card. The problem is once you go down that route, you're stuck. Christmas, birthdays, anniversaries, the list is endless; and she's got a lot of friends.

Every year the cards have to become more creative too. She has an arsenal of glue guns and glitter, rhinestones and markers; and she puts pressure on herself to improve upon previous efforts. When she's making a card, she's in lockdown.

'Can't talk,' she'll say, 'only two weeks left.'

Now that she has a boyfriend with Valentine's Day coming up, she could give him a child and it would be less work.

Giving homemade gifts is one thing, receiving them is another. My friend Nina knit me a scarf which I love and wear all the time. But there are some loops coming undone and I'm worried it will unravel.

I sent her an e-mail mentioning this. She wrote back: I did the best that I could.

Should I not have said anything? It wasn't a criticism, it was an enquiry. If someone makes you a gift, and it falls apart, then what? It's a lot of pressure.

Making something should be a magnanimous gesture with no expectation that it will be reciprocated. If I buy someone a bag, I don't want to feel bad that I didn't tan the leather.

Not everyone is willing or able to bake their own bread or grow their own vegetables. There are plenty of reasons I let people down — not being able to hand-sew a kimono shouldn't be one of them.

# Mistaken Identity

Most days I don't want to be seen. If for some reason I have to leave the house, I'll walk really fast to wherever I need to get to, head down, eyes on the pavement. Waiting for the light to change is an endurance test. I'm exposed and vulnerable. Someone could see me from across the street or walk by as I'm standing there.

When I don't look good and I run into someone, I know that as soon as they walk away the only thing they're thinking about is how I've gone downhill.

Sometimes, to make myself feel better, I'll think of a random celebrity. Gwyneth Paltrow, for instance. I've seen pictures of her in tracksuit bottoms, no make-up, hair un-brushed, and I can tell she's been caught on a day when she's not really wanting to be seen. I empathize because the last thing I'd want is someone taking my photograph and having it end up in a magazine. But the empathy doesn't last long. Because I only wish I looked like her on her worst day.

The only thing I hate more than running into someone when I don't want to be seen is being seen by someone without knowing it. The other day, I got home and there was a message on my answering machine from someone I used to go out with and hadn't seen in a while.

'I saw you,' he said, 'on Fourteenth Street. I didn't say hello.'

That was the message. If you see someone and avoid saying hello, you don't call them afterwards to tell them. What good does that do? Essentially, he's saying: 'I saw you, you didn't see me, I win.'

I called him back. 'Why didn't you say hello?' I asked. He told me he was running late for an appointment. Just then I had a horrible feeling he saw me on the way back from the gym. So I asked: 'What was I wearing?'

He told me I was wearing a sweatshirt and I felt myself sinking. There's just something so unfair about being seen on the way back from the gym, looking sweaty and gross and then finding out about it. So I did the only rational thing I could think of. I lied. 'That wasn't me,' I said.

He was silent. 'Really? It looked just like you.'

Nope, I said. I told him it wasn't me. 'Really?' He repeated. 'You weren't wearing a sweatshirt and race-walking on Fourteenth Street yesterday?'

Race-walking? Now it sounded like I should be on a boardwalk in Florida. I told him I might have been on Fourteenth Street and I might have been wearing a sweatshirt, but I certainly wasn't race-walking.

'Really?' he said, for the third time. 'Are you sure?'

'I'm sure,' I snapped. He sounded like he didn't believe me. Even though I was lying, he didn't know that.

I asked where he was and when he told me he was across the street I pointed out that if he was across the street, it's not as though he got a good look. After a few minutes of convincing him that it was a case of mistaken identity, he finally relented. Then he said: 'Okay, I accept it wasn't you. Whoever it was, she looked pretty.'

# sixteen

IF ONLY

# Bionic Boyfriend

Some people are nervous about robots. Not me. They can't come quick enough. Scientists have already developed a bionic arm, they are designing a bionic eye, and bionic legs are being tested too. So here's what they need to work on next: I need a bionic boyfriend.

He would be able to lift me up and make me feel thin no matter how much weight I gained. I could call him on the phone, say I needed to see him, and seconds later he'd be on my doorstep.

Amazing feats of strength and speed would be complemented by his ability to stay awake as long as I wanted. 'Go on, keep talking,' he'd say. 'I don't need to sleep.' Naturally, his bionic breathing would prevent him from snoring.

My bionic boyfriend would have no problem with how much I packed. Lifting three cases for a weekend away wouldn't be an issue. Never again would I miss a taxi or plane. I'd just ask him to run after it and bring it back.

His strength wouldn't just be physical, it would be emotional too. I could be as difficult as I wanted and never have to worry about him giving up on me. He'd have bionic tolerance.

I could take half an hour ordering Chinese food, he wouldn't mind. And he wouldn't be embarrassed if I cried in public.

But this got me thinking. Even better than having a bionic boyfriend whose movements are controlled by his thoughts would be to have a bionic boyfriend with movements are controlled by *my* thoughts.

The sex would be perfect. The chip in his head would receive my signal when it was exactly the right time to make a move. Or not. When it was time to back off, he'd know. No more awkward conversations about how it really wasn't his fault, I was just tired. If his bionic arm slid in the wrong direction, I'd redirect it to rub my feet. By just thinking about his movements, I'd control the pace, the degree of dexterity, not to mention his bionic endurance.

Since I'd be at the wheel, I'd be able to choose who he assisted with his powers. Helping grandma move her piano? No problem. Helping the cute ex-girlfriend move house? No way.

I could also download his vision from his bionic eye and see what he saw when I wasn't around. If he was staring at a waitress for too long, I'd know. Then I'd recalibrate him to look at a baked potato. I could go through his e-mails with him, reading every one and having him reply to any female invitations to meet for a drink with: 'No thanks, I'm too in love with Ariel.'

I suppose it could get tricky, though. If there were bionic boyfriends, there could be bionic girlfriends as well. And what if they got together? I have enough to worry about with the human competition. Another concern is an absence of vulnerability. If he couldn't experience anger, doubt and despair – where's the fun?

Plus, thinking for myself is a lot of work. Planning his moods, moves and responses means I'd have to think for two. That sounds exhausting. And what if he broke down? It's not as though I'd know how to fix him. Then I'd be back to square one and have what I've always had: another broken boyfriend.

# DNA

A recent scientific study has shown that there is a particular gene, present in two out of every five men, that makes them less likely to form a lasting monogamous relationship. Certain men are biologically predisposed to cheat. Infidelity is in their DNA.

We don't need to unravel the double helix to know men are programmed.

But if there's truly a genetic link, I would want a screening test. Think of all the time it would save me asking questions. Where were you? Who did you have lunch with? What did you order? All potential boyfriends or husbands would be screened with the sexual fidelity test ahead of time. Cotton swabs will be flying off the shelf.

Not that I'd trust the test. It won't be a 100%. They never are that accurate, so there's still room for suspicion. Precautions will have to be taken. People with fair skin need a high SPF lotion and stay out of the sun to avoid melanomas. Will men with this gene have to stay out of Las Vegas?

The downside of the test would be if a guy doesn't have the gene and cheats anyway. What does that say? Just that he's bored with you.

Even if they got tested though it wouldn't matter. A lot of women will overlook the results. They'll still find a way to make it their fault. I can hear it now: 'Maybe I mutated the gene and it kicked in.'

Other women will use it as an excuse to become more invested. They'll believe they can change him. They'll try to figure out what triggers the genetic impulse so they can block it. Since there are varying degrees of infidelity it will become a project. Is he a chronic philanderer? Or is it seasonal? Is it a one-off and he's plagued with guilt? Or is it just his way to exit a relationship. As long as there's biology involved, women will find a way to forgive.

They only conducted this test on men because the hormone it acts upon plays a larger part in their brain than it does in women's brains. So if science can prove that men have it in their DNA to be distant, can it explain why women have it in their DNA to be drawn to emotionally unavailable men?

I bet there are certain women who have an acceptance gene. It's in their DNA to tolerate bad behaviour.

One interesting point is that research has also shown that men who have the infidelity gene have a natural ability to circumvent a woman's defences and get into her apartment. The top five excuses were:

1. 'Can I come in and use the loo?' It's hard to pass that up. No, I'd rather you pee in your pants?

2. 'I'm thirsty – can I have a glass of water?' That too. He could be diabetic. Or dehydrated.

3. 'Do you mind if I leave my computer inside so I don't have to carry it around?' I don't want him to get an ache in his shoulder and need surgery.

4. 'Is it okay if I check the football results?' Saying no to that is like denying a man oxygen.

5. 'I need to sit down for a minute – on your bed.' Quickly followed by: 'I promise nothing will happen, we can just cuddle.'

What women need is a detection gene. It releases a hormone that repels untrustworthy men. If some men are programmed to cheat then women need a biological imperative to avoid them.


# Sick Men

There's nothing better than having the man you love be sick with a cold. Some of the best times I've ever had are nursing boyfriends who are unwell. They're usually bedridden, so they're in one place. I know where they are and they're reachable at all times.

They listen to me because they have no choice. I can talk about anything I want and they'll pay attention. And even if they're not paying attention, they can't move. I can get away with asking the same question twice and not have to worry about being snapped at. They don't have the energy to argue.

Also, they need me. And they're grateful. When else would retrieving a glass of water be considered a charitable act?

The perfect situation is when a man is mildly ill. A serious illness is a different matter. Ideally, he should have a slight fever – no more than 101. Any higher than that and all they do is sleep.

Even better is when they're on antibiotics. Handing him a pill makes me look useful. Being nauseous is also a plus. No matter what I feed him, I'm an angel. But he can't be nauseous to the extent that he's vomiting – that's too much work.

The best part of this situation is that, for a brief period, he's complaining more than I am. It levels the playing field. His defences are down, which makes him vulnerable; so he's sexy – but in a way that feels safe. It's hard for a man to be arrogant with a runny nose.

I know people say the true test of a relationship is going away together, but I'm not so sure. I think the true test is when he gets sick.

What constitutes 'sick' is phase one. Getting a splinter and going to A&E isn't appealing. It's annoying. A broken limb or the flu – that qualifies. As long as it's not avian flu. Or anything contagious.

Phase two is all about discovering how he handles feeling under the weather. Some men prefer to be left alone. I hate it when that happens. Other men take it too far. Being a caregiver is a choice – not an obligation. Just because you're sick, it doesn't mean I'll do your taxes.

My ideal scenario is to have my bedridden boyfriend marvel at how caring and generous I am. I will ask if he needs an extra blanket, and just then he'll realize how special I am and decide that all of my negative qualities mean nothing. So what if I'm depressed? I'm there with the Kleenex. 'You're so devoted,' he'll say. Which, under normal circumstances, would translate as 'clingy'.

'It's nothing,' I'll reply. And mention how, when I'm feeling unwell,

I'm sure he'd do the same for me. Except that when he nods in agreement, I won't believe him.

The downside is that he'll start to get better. And as soon as that happens, I know it's only a matter of time before I get demoted. Once he's back on his feet and healthy again, he'll realize that thinking I was the greatest was just a moment of weakness.

# Unavailable Male

There is a type of man in New York who's known as the unavailable male. He's always busy, too busy to get together but not too busy to devote the time to telling us why.

Unavailable male type one: this is the type my friend Heather had plans with. He's a writer. Even before their first date, he sent her a list of his day's activities, ending with how, at 6 p.m., he was going to 'dedicate a plaque to Edith Wharton, then have some Japanese food with the dean of Columbia University, and maybe a few glasses of sake, blah blah blah.'

What kind of writer writes 'blah blah blah'? In his case it means: 'I'm too busy to write down all the mundane details, but I'm really trying to impress you with my very important schedule.' That way you know you had better be grateful that he's making time for you.

Unavailable male type two: this man is unable to make plans more than five minutes in advance. I went out with one and every time I'd try to make a plan, I felt like I was asking him to set a date for the wedding.

When he did commit to getting together, it was never a time or a place – just a vague idea of when it might happen. He'd say: 'Tuesday should work.'

Then on Tuesday he'd e-mail: 'I have to meet someone for a drink in the early evening, then there's a birthday party I have to go to, so I'll call when I'm done.' Then, when he never called, the next day he'd say: 'Well, I wasn't done till 2 a.m. and didn't want to wake you up.'

I considered this to be thoughtful. But also, it was a reminder that he was capable of making some plans; just not with me.

Unavailable male type three is always busy. My friend Emily is so fed up hearing how 'busy' her dates are, she now refers to it as 'the B-word'. As in: 'He used the B-word, so I dumped him.' She runs her own business. How much busier can he be? Some men will always be busy, even if they're unemployed.

I went out with someone who, when I asked how he was, his answer never varied: 'Stressed.' I'd ring to see what he was up to. 'Not a good time, stressed.' He was stressed in the morning, stressed in the evening. Eventually I told him to call me when he wasn't so stressed. I didn't hear from him again.

Unavailable male type four will give the impression he's available, but he's not. He'll agree to go away for the weekend, but will be on the phone, looking at his watch the entire time. I challenged this once and asked: 'Somewhere else you'd rather be?' It taught me a lesson: don't ask a question if you're not ready for the answer.

Unavailable male type five is the 'tired' man. He's closely related to the 'stressed' man and 'busy' man and he'd love to go away for the weekend. But he's too tired. There's a limit to how fast you can drive a car. How about a limit on how many times a day one person is allowed to say they're tired? Over five times and you're punished. Besides, who *isn't* tired?

An interesting fact. Call a busy man at 11 p.m. and say you want to talk: he's tired. Call him at 11 p.m. and say you're in the mood for sex and he's ringing your doorbell before you've even finished the sentence.

Occasionally I'll get a call from a man I know who has a block on his phone number. What shows up on the caller ID is: unavailable.

I should take that as a sign.

# Hard Work

People always say nothing in life is worth having unless you've worked for it. Hard work pays off. Except when it comes to relationships. Or more specifically, relationships involving a woman who is 'complicated' with a strong personality.

I was having coffee with a man who was telling me about a failed relationship. He was going over it, trying to figure out what went wrong, and why, even though he loved her, he couldn't be with her. I was sympathetic. Right up until he reached a conclusion. She was, he said: 'Too much like hard work.'

What does that mean? What vaults someone into this undesirable category? The category that wipes out all other redeeming features? And once you're in it, there's no getting out. The more you try to prove you don't belong in this category, the more it reinforces you do.

I felt bad for this woman I'd never met. I wanted to call her up and tell her to move on – quickly – because unless she was willing to get a lobotomy, the relationship was definitely over. And then, just as I was about to mount a defence on her behalf, my friend said, 'You know what I mean.'

Now I was confused. Was he suggesting I could relate to what he was saying because I have been in situations with men who were hard work? Or, did he mean I could relate to his decision because I have 'hard work' stamped on my forehead? And that I have experienced my share of men giving up on me – and who can blame them?

It was the latter. 'C'mon,' he said, 'you know you're not the easiest person to be in a relationship with.'

How would he know? We've never even dated.

'You think I'm hard work?' I asked.

He gave me a look. 'You know you're difficult,' he asserted. 'It took six e-mails, three phone calls and a dozen texts to make a plan for coffee.'

That's not difficult, it's thorough. And therein lies the problem.

Everyone has their own definition of what constitutes hard work. What some people see as a headache, others see as a challenge.

If something is bothering me, I tend not to keep it inside. This might not be the way to go. Nothing says hard work more than a woman who starts conversations with: 'Can I ask you a question?' From now on, when someone asks me what I'm looking for in a man, I'm going to say: a good work ethic. That, and a high pain threshold.

# Separate Homes

If I were ever to get married, I'd insist we have separate residences. Every time I've heard about couples living in separate houses I've loved the idea. The painters Frida Kahlo and Diego Rivera had houses next to each other with a walkway that connected them. Tim Burton and Helena Bonham Carter live in adjoining flats linked by a corridor. My preference would be to live across the street from one another. That way, I could look into his window to make sure he was home.

A friend told me she is having problems with her husband, who snores. It's got so bad that she's now sleeping in the guest room. I told her she'd be better off if she were to move out entirely. I can see advantages in waking up together, but continuing to share a bathroom? Where's the romance in that?

Audrey is against living apart. She says she likes the domestic stuff that goes with cohabitation. When I asked for specifics, she said: 'It's great having someone to share the boring stuff with – doing the dishes, laundry, taking out the rubbish.'

I don't see why that should have to be forsaken. That's why God invented speed-dial. 'Honey, can you come over and help take out the rubbish?'

But Audrey also believes that figuring out how to live with someone can be really good for other parts of the relationship. I'm not so sure. How is taking out the rubbish together going to improve our sex life?

I'm convinced separate houses are the way to go. After sex we could lie together for a while, but as soon as he started to snore I'd go home – or, even better, he'd go home so I wouldn't have to get up and get dressed.

I would never have to panic if he bought a plasma TV and announced he was so excited to watch *Dirty Harry* for the fiftieth time. And think of all the nagging it would eradicate. If he wanted his things lying around, I wouldn't care. He'd have his space, I'd have mine – and it's not as though I can't keep an antiseptic hand wash at his house.

Then again, I might begin to wonder what he was up to. If we lived across the road, and I saw the lights on, I'd find it odd if he didn't answer the phone. And what if he had friends over and didn't invite me? That would be weird. I would assume I could go over anytime. Because I'd have a key.

But would that mean he'd have a key to my place? Here's what I'd like. I'd have a key to his place, but he wouldn't have a key to mine. What are the chances of that working out? Maybe I think the secret to a happy relationship is not living together, because deep down I know we'll eventually break up – and when we do, it means I won't have to find a new place to live.

# Tragedy Man

People get married for different reasons. Sometimes it's for money and status, or to start a family. Sometimes it's for the companionship or because it's what they feel they should do. I've never felt compelled to get married for any of those reasons. I've never felt I needed to get married at all.

But there's nothing like sitting alone in a hospital waiting room to make you realize you want a husband. Waiting for my father to get out of the operating theatre was agonizing. Every terrible thought went through my head. Just then it hit me: I wish I had someone to share this with.

I want that commitment. If only there was a way to marry someone who would show up during the tough times – funerals, hospital visits, disease diagnosis – whenever I'm scared and needing support. And the rest of the time, there's no obligation. We don't even have to have sex. The only requirement would be that when a loved one got sick or died, he had to hold my hand, supply tissues, and listen to me talk. And keep track of my keys.

Afterwards, when the crisis had passed, he could go away. He could live on his own, travel, lead his own life – but like a surgeon, he'd be available when tragedy struck. People might say that's what a therapist is for. But a therapist won't spend the night.

When it comes to finding a partner, people always say it's good to know what you want. But sometimes it takes a major event to realize what that is. Going through life on my own, I'm fine. Going through a crisis, I'd like someone to lean on. Someone who can help me find where I'm going and make sure I don't lose my wallet. Someone who knows what to say or, even better – knows when not to say anything at all.

I'd love nothing more than to be there for him during the tragedies in his life too. I'd feel special. The one time I got a call from a man saying, 'Get in a taxi – quickly – I need you' it was because he was in the mood to have sex. Essentially a booty call from someone too lazy to leave their own apartment. Not the kind of special I had in mind.

# Proposals

Cherie Blair revealed that while on holiday in Tuscany, Tony proposed to her while she was cleaning the toilet. 'He said, while I was on my knees, that maybe we should get married.'

Why was she cleaning the toilet on holiday? If you're with a man who doesn't mind that you're cleaning the toilet while on holiday, he might as well propose to you while you're at it.

But if you think about it, it's what he said, not where he said it. 'Maybe we should get married.' Maybe. It sounds so casual. I'm not sure how I'd want someone to propose to me, but I know what I wouldn't want. I wouldn't want him to ask before I've had my coffee. If I can't make a decision about what to wear in the morning, chances are that I'm not going to be able to decide about a lifelong commitment.

Also, if he suggests meeting at our 'spot', that could be problematic. Because I'll get it wrong. He'll be waiting somewhere, I'll be waiting somewhere else, we'll text angry messages to each other, and the next thing you know, he'll be wondering why he even bothers.

The less fanfare the better. No roses, no beaches, no string quartets, no sunsets. But most importantly, no limos. Nothing tasteful has ever come out of a stretch limousine.

Another proposal I wouldn't trust: he asks me to marry him on a plane that's about to crash. That wouldn't count. In fact, any scenario that involves imminent death – I'd have to wonder if he really meant it.

If there was a ring involved, putting the ring into anything edible is a bad idea. I'll bite down on it and crack my tooth. Either that or I'll swallow it and end up in A&E with a ruptured spleen.

I don't need a ring, but there should be something to signify the moment. Nothing says 'I haven't really thought this through' like a proposal written on a napkin. And if there's going to be a proposal while in a taxi, it would be best to cover the fare. I'm all for splitting a cab but if you're going to propose, at least deal with the tip.

I've heard of people proposing using a treasure hunt with clues, and each clue is a joke or a riddle that needs to be solved. This would not be the way to go. It would take me so long to solve the first riddle that by the time I'd reached the end, he'd have moved on and married someone else.

The circumstances surrounding the proposal are crucial too. For instance, if he's had to move out. 'Let's get married' should never follow the sentence: 'I've just been evicted.'

If it's outside, it can't be anywhere tropical because of the mosquitoes, and it can't be anywhere chilly because I'll be cold. Writing in the sky also wouldn't work. I'll be bending down to tie my shoelace at the exact moment the plane flies by, then when I look up, all I'll see is a smudge of smoke. When I say it wasn't my fault I missed it, he'll take it as an omen that we're not meant to be.

Here's a proposal I wouldn't mind. He holds a gun to his head and says: 'Marry me or else.' Now that's romantic. I'd know for sure how he felt. But then I'd wonder if maybe he was just saying that. I'd want to say no – just to test him.

# Monastic

Monasteries and convents are advertising weekend retreats as a way of enticing people into religious orders. I've been thinking I should try it and see how it goes. I need a retreat.

There are several things about life in a convent or monastery that would suit me. To begin with, the wardrobe. A baggy muumuu with bell sleeves and a hood. What's not to like? It wouldn't be that much of an adjustment either; it's similar to the bathrobe I live in now. If I were a nun, the bonnet would be even better – I'd never have to worry about how my hair looked again.

In my life now, thinking about what to wear every day fills me with panic and my wardrobe is essentially monochromatic. I'm not good at making decisions, and I don't even go out. All the time I'd save I could devote to more productive endeavours. Like sitting.

The activities of monastic life would be perfect. Monks tend to do a lot of things I enjoy – ruminating for instance. I do that all the time. And all the uncomfortable feelings I have in the outside world would be appropriate in a monastic setting. When I sit in a therapist's office anxious about feeling detached it's a bad thing. But in a monastery, it would be Zen.

None of us would wear make-up or drink and I'd love the fact that no-one, other than God, would ever have to see me naked. Never again would I have to stress about finding time for a waxing appointment.

Everyone tells me I have commitment issues but maybe that would change by committing myself to the Lord. I haven't gotten the call from Christ yet, but I don't mind waiting. I've waited for a lot less interesting men to call and they only thought they were God.

Another thing: monks and nuns go to bed early. At 9.30 everyone would be wiped out. I'd be in my element. Also I wouldn't have to do anything too strenuous or panic about social gatherings. What are the chances a monk is going to say, 'Today we're going horseback riding'? I'd never feel left out.

But then, after a week or so, I can see that there might be problems. In a convent, the nuns would ask me how I was doing and I'd say I was craving sushi. They wouldn't like that answer. They'd invite me to sit in on more prayers and I'd wonder if they were really accepting me – or just saying it. And I'd miss talking on the phone.

If I were in a monastery I'd start to get fidgety during the meditation and the monks would ask me to sit still. The chanting would get on my nerves. And when I told them the robe was itchy they'd get annoyed. We'd sit around at dinner and what would the conversation be like? 'How's the soup?' Salty.

Then, I'd ask one question too many and that would be it.

Is it possible to be expelled from a monastery? I bet it is. That can't be good karma.

Now that I think about it, I'm not sure life in a religious order would suit me as well as I'd hoped. Medieval buildings are cold. If I need three hot-water bottles in my flat with central heating, I'd have no circulation at all in an abbey. There's also the fact that I'm Jewish. Nobody wants a neurotic nun.

I can see the value of a monastic existence – but I'm not sure that I'm bypassing it right now. Having no social life, feeling disconnected, and wearing a robe all day sounds a lot like my life already. And I don't have to wake up at 5 a.m.

# Adult Adoption

I've been thinking about how great it would be if I could be adopted. I know there are plenty of unwanted children in the world, but there are a lot of unwanted grown-ups too. How old is too old to be adopted? No forms, no waiting, and I'm already paid for. Think of all the money it saves on school fees, medical expenses, and clothing allowance – although I'm still open to being taken shopping.

I know Brad and Angelina are looking but I'm not sure they would be the best family for me. Maybe as a foster family – for a few months – so that I could travel on the private plane to Africa and Asia and stay in a few châteaux. But then I'd have to move on. Their lifestyle is too exciting.

I'm looking for a family with low expectations. A family where there's no pressure to be joyful. Especially around the holidays or in social situations.

I wouldn't mind having a brother or sister but they couldn't be too successful otherwise I'd get depressed. Or, they could be successful but in something completely non-threatening while at the same time useful. Maybe a sibling who was a leading neurologist. Or head of upgrades at Virgin Atlantic.

I'd also need a family that doesn't like picnics or parades. I'm not good at small talk and I don't enjoy outings. If my new family had a beach house, that would be okay as long as they didn't expect me to go there in August and let me use it in the winter when no-one was around.

If it was a family that was adventurous, I could adapt. If they wanted to go sailing, I'd wave to them from the shore. Everyone needs that person in the family – the one who stays home. I'd be the cheerleader from the sidelines. Unless it was too cold. Or too hot. And there'd be no pompoms.

I told my friend Sophie I'd like to be adopted and she asked what I had to offer a family. I hadn't thought of that.

Discussions on existential angst? Free of charge. If they were ever worried about a disease or an infection, I'd be happy to Google the symptoms or speculate with them about a diagnosis.

For an older childless couple, I'd be better than nothing. I could check in every so often and make sure no-one had died. Then if one of them died, I could help with the funeral.

'I wouldn't mind being adopted,' Sophie said. 'Provided it was by a family of underachievers. That way they'd think I was brilliant no matter what I accomplished.'

When I asked her what she had to offer a family she replied without hesitation.

'Not much.'

But then, after thinking it through, she said, 'I could offer them a place to stay in New York.' As long as it was a small family and they weren't too tall. She has low ceilings.

I pointed out that at least she knows how to knit. That's a coveted skill. Why would anyone adopt a baby when they could get a quiet forty-year-old who has a microwave oven and an ability to make scarves?

If a granny adopted her, they could sit at home and knit all day. With the cats.

The upside to being adopted as an adult is that my new parents wouldn't have to worry about whether or not I'd want to meet my biological parents. The downside for me would be I could see myself becoming really attached and then they'd decide it's not working out. And ask for a refund.

At least they wouldn't have had to cover the cost of a wedding.

# Life for Sale

A man in Australia put his life up for sale on eBay. His marriage ended and he decided that he wanted a fresh start. He auctioned his life in a 'package deal'. Everything from his house, his car, furniture, clothes, friends – even a two week trial run at his job. Who would want to work at a rug shop in Perth and wear someone else's smelly T-shirts? And how much can his friends be worth if he's willing to tack them on too?

If I were to put my life up for sale, I wonder how much it would go for.

People would definitely bid on my life: a small but centrally located apartment, brand new Henry vacuum cleaner, and a medicine cabinet fully stocked with antibiotic and homeopathic remedies – gels, creams, lotions. There would never be a need to go to a pharmacy again. It would be good value.

There's also a gym membership, writing for a prestigious newspaper, loads of air-miles, a handful of good friends in both New York and London and I'd throw in my shoes (size 36) all of which are in mint condition since I never go out.

They'd get regular appointments to the dentist, gastroenterologist, gynaecologist, urologist, dermatologist and most important: the private e-mail address of Dr Samuelson, my GP. Who else could offer this kind of access? It's like having a twenty-four-hour hotline to the *New England Journal of Medicine*.

Another good bargain in buying my life would be getting my friends. Although, after the buyer actually purchased my life they'd realize my friends are very busy and it takes months of planning to see each other.

There are a couple family 'situations' that they'd have to deal with. And some debt. I'd have to make sure I said: No Returns.

On the Australian's website there is a welcoming message from his friends ending with the sign-off: Happy Bidding. On the website describing the contents of my life in the Friends section they'd leave the message: Don't call us, we'll call you.

Then again, my friends might jump at the chance to make plans with whoever took over my life because they'd probably be a lot more fun to hang out with than me.

That would be awful. What if the person who bought my life really enjoyed it and had a great time? All of the fun, none of the anxiety. Waking up eager to start the day – making plans to do things like go to concerts and parties – saying yes to invitations instead of no?

I can see it now. I'd watch someone else enjoying my life after they bought it and then suddenly I'd want it back. Only then it would be too late. Just as I realized its value, I wouldn't be able to afford it.

# Companion for Hire

Here's a new job: Sober Companion. Also known as Clean-Living Assistant. Someone gets paid $2000 a day to make sure a celebrity doesn't drink and prevents them from relapsing. They're on retainer to follow the person around the clock making sure they stay clean.

That's a job I have some qualifications for. I don't drink and I'm very clean. Also, I'm good at sleeping late and talking. The only problem is I'd have to talk to celebrities. And celebrities love to sit in the sun. That would be tough.

I guess it would depend on who hired me. If George Clooney were looking, I'd offer to work weekends. But if Paris Hilton hired me, I'd last 10 seconds. As soon as she said she was going to tan, I'd go home.

Another variation on the same job is an Eating Coach. The job consists of following the celebrity around on errands and making sure they don't skip meals or think a piece of gum counts as lunch. I'd be good at that too. 'Eat something or you'll end up with kidney failure.' Who wouldn't listen to that?

My friend Sophie says her dream would be to hire a Reality Check Companion. Whenever a man came into the picture she would be on duty. The job would be to point out why this man was an unsuitable romantic interest and refer to things he was doing or saying that were red flags.

Then no matter how much Sophie tried to rationalize it could work, the Reality Check Companion would refute it. In dire circumstances she would bring up the exes as a cautionary tale. Emotionally, the working conditions might be stressful. Especially around Valentine's Day. On the other hand, the job would be part-time. No-one wants full-time reality.

I'm not sure what sort of assistant I would hire. Maybe a Wellness Companion. Every time I thought I had something he or she would point out why it probably wasn't what I thought it was.

I'd say: 'How do you know?' They'd say: 'I don't know for sure' and I'd say: 'But that's your job — to know for SURE.'

I can't see anyone qualified for that position. Except a doctor. But if you've gone to the trouble to get a medical degree chances are you don't want to follow me around while I do my chores.

# Amnesia

I read about a man who woke up at the end of the subway line in Coney Island with no idea who he was. At the police precinct, they thought he was on drugs, but doctors diagnosed amnesia. It was a rare form – the kind you see in movies – where you remember places and how to eat and function, but have no memory of your life. Your entire past, friends and relatives – all erased.

That sounds pretty good. I wish I could get amnesia. I can't think of anything I'd like more than to wake up one day with no recollection of my life or who I am. Especially if it was the kind of amnesia that comes out of the blue. Not the kind that comes from a car accident or blunt instrument to the head, or anything remotely painful. I wouldn't want to wake up in hospital. I'd prefer to wake up in Notting Hill. With no ID and a suitcase full of money.

Here's why I'd love to have amnesia: I'd forget who I was and wouldn't have the grief of hating myself. I'd forget I'm turned off by certain random expressions and this would open up a whole new world of dating possibilities.

I'd forget I had anxiety about being seen naked, and I could have sex with the lights on – or during the day. Options would open up. I'd forget that when a man says 'awesome' it's an automatic deal-breaker.

I would never have to worry about running into someone I didn't want to see, like an ex I needed to avoid because he didn't call. I'd forget that I was a procrastinator and I'd clean my apartment. I'd forget I hate the gym, so I'd work out. Then I'd get invited to parties. And I'd forget I was anti-social, so I'd go.

I keep thinking about the guy who developed amnesia. It's got to be tough on his girlfriend and family. That's why I'm a perfect candidate; if I had amnesia, most people in my life would be relieved.

# Plan B

I heard about a very successful hedge-fund manager who ripped off millions from his clients, got caught, faked his suicide, and went on the lam. He used to live in a building owned by Donald Trump and now he was living in a campervan. That was his Plan B. Going to jail sounds more appealing.

Where would you go and what would you do if it all came crashing down? Sometimes I like to think about this. If everything in my life exploded, what kind of life would I have if I had to start over?

Most people have an escape fantasy — a life they dream about and aspire to. I haven't been able to come up with anything meaningful.

Sophie has it all planned out. If her life imploded she would buy a house in a small New England town and get a job working at the local diner as a waitress.

'I'd be the woman with a mysterious past,' she says. 'An enigma.'

I like the idea of being an enigma. But I don't think it would last more than a week. One day someone would ask, casually, 'What's your story?' Five hours later I'd still be talking. Either that or I'd say I can't get into it because I'm an enigma.

If you're going to be an enigma, it's not something that can be announced.

Also in Sophie's plan is a romance. She'd fall in love with the vet who looks like Sam Shepard. Her plan sounds a lot like the plot from Baby Boom. Only without the baby.

Whenever women talk about their fallback plan it always involves a small town with an ordinary guy who just happens to look like Sam Shepard. But I've been to a lot of these small towns and there's never anyone who looks anything like him. If there is, he's slumped on the street corner with an empty wine bottle in a brown paper bag.

But at least Sophie has a plan. Maybe I could rent a room in the house that she's buying. The problem with that is, what if my life implodes first? I'd have to wait around for her life to fall apart and that could take a while. She would continue to have a successful life and I'd be stuck waiting. I'd become resentful. It would put a strain on the friendship.

I need a plan of my own.

I asked my friend Simon, what would happen if his life fell apart. What's his Plan B? 'My life right now isn't exactly Plan A,' he said. 'I'm already on Plan G.'

But then, after pressing him for a while, he came up with having a fish farm. I couldn't follow all the details but it was definitely different to being a fisherman and it wouldn't involve having to listen to a shipping forecast. When I think of fish, I think of water. When I think of a farm, I think of land. I was confused. And not interested. It sounded wet and cold.

Another friend mentioned he'd become a trucker. Why? What's romantic about sitting for twelve hours a day in traffic, drinking stale coffee from petrol stations and getting haemorrhoids? My Plan B needs a place where I can pee without wiping the toilet seat first.

My friend Heather, who dates a lot, told me her fallback plan was to travel around the world with her husband. 'You could join us,' she said. That sounded good. But then I remembered she's not married. And where would the money come from? I pointed out that a fallback plan is for if life comes crashing down. What she was describing was hitting the jackpot.

I'm very specific. I can't imagine my fallback plan would involve being anywhere too remote. But I like sheep. I wouldn't mind being a shepherd if it could happen somewhere close to Manhattan or London. The downside is I have a feeling a shepherd's hours wouldn't work. I'd oversleep and lose my flock.

The fact is, I don't have a Plan B. It's beginning to worry me. If everything fell apart tomorrow I'd probably end up hiding out in my apartment ruminating on where it all went wrong.

Then again, that would keep me busy for the next ten years. And it's familiar.

## You Never Know

People love to say 'So much can change in a year.' But they always use that to point to good things and new beginnings. Meeting the love of your life, getting married, pregnant, moving, promoted, and so on. But what about when so much can change – for the worse? This time next year – who knows what can happen.

My father has a great philosophy. He says, 'You are the same person you've always been. That doesn't change.'

Yeah. That's a problem.

# ACKNOWLEDGEMENTS

There are many things I am grateful for. One of them is having met Laura Barber and been lucky enough to have her as the editor of this book. There has not once, in this entire process, been a moment where I've felt she doesn't get it. From the moment this all began, her insight and judgement has been consistently, infinitely helpful. I am grateful, too, for her optimism and everyday thoughtfulness. Whether it's finding a way around track changes or being endlessly gracious and kind when answering questions she's probably answered already, twice, her patience has spoiled me for life.

And what other editor on earth would begin a phone call with, 'Everything okay? No fire in the kitchen?'

Thank you to all the Portobellos for their time, attention, advice and help. I could not have asked for more supportive publishers.

My agent Tif Loehnis at Janklow & Nesbit has been a longtime supporter, and heartfelt thanks go to her and to Will Francis, whose efforts on my behalf are deeply appreciated.

At the *Guardian*, thank you to Ian Katz for giving 'Half Empty' a chance, and to Esther Addley, for editing it in the spirit in which it was intended.

Special thanks go to everyone at the *Sunday Times Magazine*. To those in particular who worked on 'Cassandra' in print: Mel Bradman, for her careful eye and Cathy Galvin, whose sensibility is darker than it appears and whose sensitivity and warmth meant – and continue to mean – the world to me.

At Times Online, I am indebted to Rhian Lewis and Lisa Zanardo, who have worked tirelessly to place and promote the column. This is greatly appreciated – especially given that they're often working at 3 a.m.

I am fortunate to have such a great team at Harper Perennial. Everyone has been extremely enthusiastic, which has made 'Cassandra''s journey to the US a pleasure. Carrie Kania, Jennifer Pooley, and others: thank you.

To my friends – I won't name you in case it's embarrassing but if you've ever heard me say: 'Can I ask you a question – it's for this week's column?', then I am referring to you. Liza P., Laura B., Katie Z., Robin P., Oliver B., Maggie M., Tamara R., Kim S. and many more – thank you for sharing your experiences, which have not only given me invaluable material, but even more important, made me feel less alone.

And lastly, this book would not exist without Robin Morgan, my editor, mentor and friend at the *Sunday Times Magazine*, whose unwavering support brought 'Cassandra' to the page, the internet, and then to Laura Barber. Thank you for caring. And for championing me and believing in me, no matter what.

# INDEX

hypochondria 223–4
melanoma 160–1, 218
male illness 262–3
Raynaud's disease 222
rosacea 33, 73
STDs 53
toothache 215–6
wandering eye 137
Wegener's disease 108
*see also* dentistry, medicine cabinet, pharmacy, sanitation
insurance nightmare 222–4
Italy 9, 16, 197, 235–8

Jane 122
jealousy 48, 203
jet lag 11
Jewishness 24, 136, 181, 273
Jon 28

Kate Moss 60, 102

LA 33, 35, 50–1, 99, 115, 241–3
lamenting 13–14
laziness 67, 68, 93
Len Rosenberg 197–8
Leve's Law 176
listening (or not) 97–8, 113, 189, 210
Linda 35, 99
Liza 9, 15, 23, 26, 30, 34, 40, 41, 48, 62, 64, 65, 66–7, 70, 81, 82, 84, 92–3, 106, 125, 136–9, 147–9, 154,

160, 199, 208, 217, 220, 225, 235–8, 251, 253
London 80, 89, 99–101, 115, 117, 118, 216, 219, 221–2, 247, 249–51
Lori 11, 31, 87, 103–3, 134, 178, 200
loser
boyfriend 209–10
friend 21–22
*see also* nobody (being a)
love 206–7

Mallery 105–6
mani-pedi 12, 123
marriage 66, 133–5, 207–9, 210, 218–19, 267–71
medicine cabinet 43, 182, 275
memory 16, 23
millionaire 200–1
mosquitoes 42, 237, 271
mother 16
mushrooms 187
mystery (air of) 8, 62–4, 279

nap politics 10–11
National Peanut Butter Day 60
neighbours 13, 79, 85, 102, 156, 208–9, 250
nervous breakdown 39
neurosis 91–2, 119, 149
New Year's Eve 102–3
New York 17–18, 42–4, 90, 99–101, 103, 105–6, 114–18, 122, 216, 220, 274